CURIOUS BONES
Mary Anning and the Birth of Paleontology

CURIOUS BONES
Mary Anning and the
Birth of Paleontology

Thomas W. Goodhue

620 South Elm Street, Suite 223
Greensboro, North Carolina   27406
http://www.morganreynolds.com

CURIOUS BONES: MARY ANNING AND THE
BIRTH OF PALEONTOLOGY

Copyright © 2002 by Thomas W. Goodhue

Library of Congress Cataloging-in-Publication Data

Goodhue, Thomas W.
  Curious bones : Mary Anning and the birth of paleontology / Thomas W.
Goodhue.-- 1st ed.
    p. cm.
Summary: Recounts the life and work of Mary Anning, who collected
fossils throughout her life and made major discoveries in paleontology
when that branch of science was first emerging.
Includes bibliographical references and index.
  ISBN 1-883846-93-5 (lib. bdg. : alk. paper)
  1. Anning, Mary, 1799-1847--Juvenile literature. 2.
Paleontologists--England--Biography--Juvenile literature. [1. Anning,
Mary, 1799-1847. 2. Paleontologists. 3. Scientists. 4. Women--Biography.
5. Fossils.] I. Title.
  QE707.A56 G66 2002
  560'.92--dc21

2002008540

Printed in the United States of America
First Edition

To Hugh Torrens
for all his help with this book.

Contents

Mary Anning, pictured here with her handpick and faithful dog.
(Courtesy of the Natural History Museum, London)

Chapter One

The Girl on the Cliff

As a child growing up in England during the early 1800s, Mary Anning would walk with her father along the shore near their home in Lyme Regis, a town on the English Channel in the southwest corner of the county of Dorset. Mary and her father were looking for the telltale signs of buried "curiosities," or odd-looking bits of rock. Spotting a large, round lump, Richard Anning would scramble up the steep cliff. Setting down his basket, he would hammer a chisel into the crumbling wall, prying away loose spiral-shaped stones called "snake stones" or "Ammon's horns." Some of the Anning's neighbors believed these stones had magical powers.

A cabinetmaker, Richard Anning used part of his carpentry shop for cleaning and displaying these curiosities that he and Mary found on the beach. In warm weather he placed some on a round table in front of an open window.

As Mary grew older, she took over her father's curiosity business, and the tourists and scholars who visited Lyme Regis each summer would buy an Ammon's horn from her table. She also sold "verteberries," which looked like giant backbones or vertebrae, and "sea lilies," which looked like stone flowers. There were "scuttle," or primitive cuttlefish, and long "thunderstones" or "arrowheads" as well.

Others had collected fossils before Mary Anning, including at least one female geologist, Etheldred Benett (whose collection of fossil sea shells is now at the Academy of Natural Sciences of Philadelphia). Mary, though, would become one of the first people in the world to recognize the scientific importance of her finds and to dedicate her life to fossil hunting. She did this at a time when the new science of paleontology, the study of extinct and fossil animals and plants, was emerging.

Mary Anning was born at home on May 21, 1799, the year George Washington died. Richard and Molly Anning had at least ten children, but only Mary and her older brother Joseph survived childhood. Death claimed four of Mary's five older brothers and sisters and at least three who came after her. During the nineteenth century, nearly half of the children in England died before the age of five. Although the population of the nation was far smaller than now, Lyme Regis was already badly overcrowded. Public health was improving elsewhere in England, but in crowded Lyme Regis, infants died at an appalling rate, mostly from smallpox or measles.

Mary was named both after her mother Molly (whose given name was Mary) and an older sister, who was killed tragically during Christmas 1798. Only four years old, this Mary was playing with Joseph in a room where

Mary and her father found "curiosities," such as ammonites, in the cliffs near their home in Lyme Regis. *(Reproduced from William Buckland's* Geology and Minerolgy, *London, 1836.)*

Richard had piled some wood shavings. Molly had left them alone for about five minutes, and during this brief time the children threw some shavings into the fireplace. The little girl's clothes caught fire and she was dreadfully burnt right before her brother's eyes.

Richard and Molly Anning hoped their new Mary, born only five months later, would survive childhood. Soon after her birth, famine swept the land. England and France, which had recently come under the control of Napoleon Bonaparte, were locked in a long war. French warships blockaded the coast of England, slowing imports and driving up the cost of food. Within a year the price of grain nearly doubled. Poor and work-ing-class people now spent half their income on bread.

A mob gathered in Lyme in March 1800, determined to prevent price gouging and mass starvation. Led by Richard Anning, they attacked estates and a mill, dam-aging so much that the royal government called out troops to put down the riot. One leader of the mob was put on trial but soon released because no one in Lyme would testify against him.

On August 19, 1800, the Anning's nurse, Mrs. Haskins, took fifteen-month-old Mary to an outdoor horse show. Mary had been weak and sickly, and Mrs. Haskins thought the fresh air might benefit her. A large crowd gathered on this hot, humid day. At a quarter to five in the afternoon, storm clouds blew in quickly, and rain began to pour. Many spectators went home, but Mrs. Haskins huddled with Mary in her arms under a tall elm tree, along with two teenage girls.

Suddenly a bolt of lightning shot through the sky and struck the elm, followed by the most awful clap of thunder anyone in the town had ever heard. There was a moment of stunned silence before a man cried out and pointed beneath the tree where Mrs. Haskins and the teenagers lay dead.

Mary's tiny body was carried home. Even before her horrified parents saw her, they smelled the stench of burnt hair and flesh. Thomas Carpenter, the local physi-cian, thought she might still be alive. A neighbor suggested soaking her body in warm water. Richard and Molly did this, praying she would open her eyes. Ever so slowly, little Mary woke up. The crowd gathered around her cheered wildly. Dr. Carpenter declared it a miracle.

One calamity after another occurred throughout Mary's childhood. In 1802, John Cruikshanks, who fossil hunted with Richard, was so discouraged by his money problems that he killed himself by leaping from Gun Cliff, right in front of the Anning home. The next year, the "Great Fire of Lyme," started by a single candle in the attic of Crossman's Bakery, burned down a cloth factory and forty-two houses. The Annings' home was spared.

When Mary was a baby, about 1,450 people lived in Lyme. The Anning home was located where the little River Lym meets the sea. It sat on top of a huge, stone sea wall called Gun Cliff, after the large cannons that lined it. The timber frame house had wooden paneling, three windows facing Cockmoil Square, and many small windows facing the sea. Below, a rough walkway called the Marine Parade spanned the River Lym down to the "Cobb," a 600-foot-long stone, gracefully curved jetty that created one of the oldest artificial harbors in England. The Cobb (which means "rounded island" in the local dialect) was probably built in the thirteenth century. One of the earliest extant ship charters, dating from 1322, is for a Lyme ship called *Our Lady of Lim*.

Two warships sailed from Lyme to join the fleet of small English boats that sailed out to confront the mighty Spanish Armada in 1588, saving the nation from invasion. In 1685, the Duke of Monmouth landed next to the Cobb in an ill-fated attempt to overthrow his uncle, King James II.

From her window, Mary could see the beach, the Cobb, and Pinhay Bay. In summer, the beach was crowded with "bathing machines" (changing rooms on wheels) and hundreds of swimmers. In winter, Mary watched waves crash against the ancient Cobb.

The Annings lived in a nice house, but it was located in an undesirable location next to the Cockmoil—the jail. Outside their door was the narrow, dangerous corner where the roads to Charmouth and Uplyme met at the bottom of two steep hills. With its high number of accidents, it was considered to be one of the worst intersections in England.

Most people did not want to live on the coast. While Mary was still a young child, a fierce storm drove huge waves up the River Lym and straight into her home, washing away the whole first floor. Mary and her family had to crawl out of an upstairs bedroom window to be rescued.

Lyme Regis, though, was one of the best places in the world to find curiosities, and the storms that pounded Lyme continued to uncover new ones. Since the sixteenth century, people had called anything dug from out of the earth—including minerals and metals— "fossils," from the Latin word for "dug up from the ground." By 1800, Georges Cuvier and Jean-Baptiste Lamarck in Paris began to study plant and animal fossils scientifically. At the same time, the new science of geology— the study of the physical structure and substance of the earth—gained interest among the educated, although at that time nearly all geologists were amateurs. When Mary first went to the beach with her father, Richard, scholars defined only those preserved objects that once were part of living things as fossils.

The Annings were "Dissenters" who disagreed with the Anglican Church or Church of England (called the Episcopal Church in the United States). They belonged to the Independent Chapel on Coombe Street, whose members were beginning to call themselves "Congregationalists."

There were many Protestant Christians in Lyme who did not "conform" to the established Church of England, but it still took courage to be one of them. "Nonconformists" could not be married legally in their own church. They were barred from English universities and from most professions. They could not become military officers or government officials. Unless they were quite poor, they had to pay taxes to the Anglican Church, no matter how much they opposed it. When they died, Dissenters often were buried without a funeral because the Church of England owned all graveyards.

Even among Nonconformists, Richard went his own way. He scandalized them by gathering curiosities on Good Friday and other church holidays. He raised eyebrows, too, by taking Mary with him. It was bad enough to most people that he let Joseph clamber up the slippery slopes with him. Mary was three years younger than her brother, and she was a girl. A rising tide might trap them along the shore. Rain turned Black Ven, the 400-foot-high cliff east of Lyme, into a dangerous mass of sticky clay. Rockslides could crush them without warning. Richard was nearly killed himself in an avalanche at the Church Cliffs near the center of town. But no matter what others thought, Mary loved this dirty, dangerous work.

Why was she so eager to do what frightened others? Did she love being with her father so much that she overcame her fears? Had she heard so often about how she escaped death by lightning that she felt lucky? Mary's own mother often said the girl was "a history and a mystery."

Whatever the reason, Mary loved being with her father both on the cliffs and in his workshop. She watched how he dealt with rich customers. He taught her how to increase the value of a fossil by clearing away the rock that surrounded it by using a needle and a tiny brush.

Richard showed her how to saw through an Ammon's horn, splitting the coiled seashell neatly in half to reveal the spiral chamber inside. He taught her how to polish each surface until its red and green hues shone beautifully, like mother of pearl. Finally, Richard taught Mary how to make fine wooden boxes to display their best finds. He fashioned a handpick for Mary that she used to probe the cliffs.

Most children in Dorset during the early 1800s had little education, but sometime near her eighth birthday, Mary Anning began attending the Dissenters' Sunday school in Lyme. The school mostly taught reading and writing, rather than religion. Unlike other schools, boys and girls studied together in the same classroom.

As Mary learned to read, Joseph Anning gave his younger sister a bound volume of the *Dissenters' Theological Magazine and Review*. All books and magazines were expensive in those days, and Mary kept the book her entire life. Her pastor, the Reverend James Wheaton, published several articles in the *Review*, many about people the Annings knew. One essay insisted God created the universe in six days, and another urged Dissenters to study geology. Many people believed that studying Earth's layers could help prove the Biblical account of Creation.

In the *Review,* Mary read an obituary for "Martha Lock of Lyme Regis, Dorset, who died October 23, 1800, aged 16 years and 5 months." By the end of the eighteenth century, descriptions of brave deathbed farewells had become popular in England. Accounts of "successful" deaths, including those of children whose last words testified to their faith and courage, were published in children's magazines.

Mary also read demands that Dissenters be allowed to hold weddings in their own chapels and arguments in favor of abolishing executions, at a time when few questioned capital punishment. Mary found an obituary her pastor had written for a Dissenter from Lyme that reflects their faith's belief in a benevolent God:

> In all my troubles thou art nigh,
> Thou sympathizing Friend!
> Thou sees my pain, and from on high,
> Dost consolation send.

This book, and Wheaton's sermons, which Mary heard as often as three times each Sunday, encouraged her to see herself as a person of value, someone with important work to do.

Soon the Annings would need all the spiritual consolation they could find. In 1807, Richard Anning set out in the darkness for Charmouth. Taking a shortcut, he tumbled over the cliff at Black Ven and fell more than 100 feet. Badly injured, he never fully recovered. He continued working when he could, but before long his body was racked by terrible fits of coughing, and he began spitting up blood. On November 5, 1810, Richard died.

The loss of the chief breadwinner left the family

without income and burdened with debt. They owed £120, much more than the family made in a year, and Richard's death did not cancel his debts. In fact, his family could be thrown in jail for not repaying them. Molly and her children had to pay rent because they did not own their home. The Annings went on "parish relief," or welfare, within three weeks of Richard's death. This public assistance was just barely enough to prevent starvation. Parish relief averaged only three shillings a week—less than half of the bare minimum needed to scrape out a living. The Annings received this charity for several years.

Mary missed her father so badly that she could not bear to visit the cliffs where they had worked. Joseph, now an apprentice in an upholstery shop, was seldom home. Their mother, Molly, was so depressed by her husband's death that she paid little attention to Mary.

One day, ignored by her grieving mother, Mary wandered alone to the shore. It was the first time she had been back since her father's death. There she picked up an Ammon's horn. On her way home she showed it to a lady she met. The woman liked it and bought it. Convinced she could earn money by selling fossils, Mary decided that moment to start fossil hunting again.

Molly Anning had sometimes complained that Richard's fossil hunting was a waste of time. Mary's sales would soon change her mother's mind about this line of work.

Chapter Two

A New Vocation

Mary's family needed every penny she could earn finding fossils. Dorset was already a desperately poor county, and England's long warfare with the French Emperor Napoleon had made it even poorer. England's new war with the United States in 1812 ended trade with the former colonies for several years, making the local economy even worse. Throughout the early 1810s, food was in short supply and prices soared. Jobs disappeared in the local industries of shipbuilding, lacemaking, and handweaving. Sailing ships had become so large that they could not fit into the harbor, and the once-busy shipyards near the Cobb went out of business. Handweaving and lacemaking, for which the town had been famous, vanished too, as steam-powered looms in cities further north made cloth and lace faster and more cheaply than it could be made by hand.

The early years of the nineteenth century were a transitional period. A new, industrial economy superceded the typical agrarian lifestyle. Old jobs disappeared and new fortunes were made in textiles, mining, and other industries. One development of this change was the growth of a wealthier middle class with more leisure time who began to flock each summer to seaside resorts. Lyme Regis soon became known as an inexpensive vacation spot. It boasted a beautiful bay, sturdy little inns, a sheltered harbor, a mild climate, and ample sunshine.

Mary and her mother began selling curiosities to the tourists, just as Richard had done. Once a week, as a coach pulled up to the Three Cups Inn, a block from the Anning home, and unloaded visitors from Bristol, London, and other cities, the Annings set out a round table filled with ancient shells and bones outside the open window of the carpentry shop.

In the early 1800s, most women spent their lives feeding, clothing, and cleaning up after their own large families, or caring for someone else's. Mary and Molly were the first mother and daughter fossil-selling team.

Mary was responsible for finding the fossils. Besides walking the shoreline, Mary frequently visited the "stoneboatmen" in the local rock quarries. Limestone for buildings, harbor construction, and ballast in sailing ships was cut from sea ledges along the shore, just below the high tide line. Quarry workers often uncovered curiosities; thus Mary checked regularly to see what they had found.

Even more often, Mary woke up early in the morning and searched the shore herself. As rain seeped through the clay and limestone in the cliffs, great chunks of rock slid into the ocean, uncovering layers of earth, stone, and fossils. Once a ninety-foot section of the Church Cliffs collapsed. Another time a large section of the road to Charmouth tumbled to the shore. One local farmer said, "All this land, 'tis in love with the sea."

The shoreline was a peaceful place to work—and a lonely one. Sea gulls and curlews flew overhead. Cormorants dived for fish, and oyster catchers pried open shellfish with their long, blade-like bills. Mary saw herons and rock pipits in the winter; in the fall, colorful goldfinches darted between the bushes on the cliffs.

Sometimes there were smugglers on the beach, but these "criminals" posed little danger. Most people opposed the high taxes on imported goods and were happy to buy smuggled tea or liquor. Except for smugglers, and sometimes someone swimming or fishing, Mary was usually alone while she worked.

Mary was twelve years old near the end of 1811, when her brother Joseph found a very odd-looking fossil embedded in rock that had fallen on the shore. It was four feet long and shaped like a lizard's skull, with room for huge eyes and two hundred teeth. Joseph thought it might be a crocodile head. He hired two men to dig it out and carry it home. Too busy in the upholstery shop to do much searching himself, he asked his

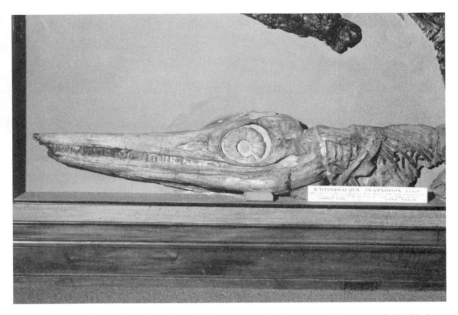

Mary and Joseph had found an *Ichthyosaurus*, the first major discovery of this kind of fossil. *(© The Natural History Museum, London)*

sister to look for the rest of the skeleton. Knowing how badly her family needed money, Mary was determined to find it.

The place where Joseph found the fossilized head was far from home—near Charmouth—and it could have fallen from anywhere in the cliff. Mary searched for nearly a year before she found almost all the bones of a huge beast, one hundred feet below the top of the cliff and thirty feet above the shore. Digging out the surrounding rock, she traced the outline of a monster seventeen feet long. The Annings hired some workers who carefully chipped away the biggest chunks of surrounding rock. Then they lifted the heavy skeleton out of the cliffside and carried it to the Anning's cellar.

The creature had flippers like a dolphin, teeth like a crocodile, and a pointed snout like a swordfish. Its backbone was like the spine of a fish, but the chest resembled a lizard's.

Henry Hoste Henley, lord of the Manor of Colway and an avid fossil collector, bought the skeleton for twenty-three pounds. He then gave it to William Bullock's London Museum of Stuffed Animals, the nearest place to a natural history museum in England. Mary and Joseph's discovery quickly became one of the most popular exhibits. Scholars and ordinary people came from near and far to pay a shilling to have a look.

The museum, and at least one newspaper, called the skeleton a crocodile. Some thought it was a fish. Dr. Carpenter, the Lyme doctor and druggist, called it a "lizard porpoise." In fact, it was so different from any known creature that no one knew how to determine what it was.

Although the discovery created a sensation, neither fame nor fortune came to the Annings. While twenty-three pounds would feed the family for six months, and was much more than Richard ever received for a curiosity, the fossil soon resold for twice as much as the Annings had received. Squire Henley owned the land above the cliff where Mary found it (and nearly all the land around Lyme), so they may have had no choice but to sell it to him—no matter what he offered. The Annings remained on parish relief, and Mary continued looking for curiosities.

During the 1810s, rich landowners feared that the poor might rise up against them in revolt—as had occurred in France during the French Revolution two decades before. The British government, still controlled by the landed aristocracy, outlawed all public meetings and all labor unions. Even those calling for moderate political change could be jailed. Peasants and most workers could not vote—and neither could half the middle class.

Public support for the government and the royal family declined during the decade. King George III suffered recurrent fits of madness and was declared incurably insane in 1811. His disreputable son, George IV, ruled for the next nine years as prince regent. His behavior created a series of scandals.

Mary's work led her into a friendship with three well-to-do sisters. Mary, Margaret, and Elizabeth Philpot moved to Lyme Regis when Mary Anning was still a young girl, and despite their differences in wealth and age (the youngest, Elizabeth, was nineteen years older than Anning), they became good friends. Mary Anning and the Philpots helped each other learn more about fossilized creatures. Elizabeth became particularly fond of collecting fossils, and she and her sisters had a large display of them in fancy cabinets in their home on

Silver Street. By the time Mary Anning was a teenager, she and the Philpots were guiding scientists along the shore. Their friendship and the company of those they led helped make their work less lonely.

Mary often guided William Buckland, a fellow at Oxford University. Buckland was a clergyman, fifteen years older than Mary, who grew up in nearby Axminster and spent vacations in Lyme. His father, the Reverend Charles Buckland, encouraged his son's curiosity about natural history. Though blind the last twenty years of his life, Charles Buckland took young William to explore local rock quarries. William described the fossil shells that his father could only touch. It was not unusual to be both a preacher and a scientist (a term that was not created until the 1840s), as geology was then seen as a romantic and a "spiritual" science. Pious pastors were among the first to learn how to read the history of the earth in layers of stone. This attitude of openness among many in the clergy would soon begin to change, as geological discoveries began to provide questions about many of their assumptions.

Mary also led Henry de la Beche, who, like Mary, had lost his father as a child. Henry moved with his mother to Aveline House, just up Broad Street from the Anning home, in 1812. Sixteen years old, he came to Lyme Regis after being thrown out of the Royal Military College for insubordination. Both Mary Anning and Dr. Carpenter encouraged his interest in fossils,

Mary supplied geologist William Buckland, here in 1832, with fossils and guided him on the cliffs near Lyme Regis. *(Courtesy of the Natural History Museum, London)*

and Henry decided to study geology, a rare choice for a rich young gentleman.

William, Henry, and Mary spent many happy hours together scouring the cliffs. They became life-long friends. Perhaps this trio was bound together by grief as well as scientific curiosity: William lost his mother in 1812; Mary and Henry lost their fathers at an early age. Whatever the source of their mutual attraction, they became an efficient team. Mary found fossils, Henry drew them, and William wrote about them for other scholars.

In 1813, a woman named Mrs. Stock, who had hired the fourteen-year-old to run errands, gave Mary the first geology book she ever owned. Soon Mary was borrowing and reading everything she could find about the new science of geology and the even-newer science of paleontology.

Finding science books was not easy. Lyme had two small libraries that rented books for a small fee, but neither had many scientific books. What Mary learned from the few geology journals and books she did read convinced her that the effort to find them was worthwhile. She learned about the types of rock formed when mud or clay settles rapidly, such as sandstone, shale, or mudstone. She discovered that limestone, which often contained fossils, is formed out of sea shells, algae, and coral. She also learned that sandstone seldom contains fossils. Mary read that the Dorset coast was a place where creatures were likely to have been preserved because the "blue lias" runs along the coast for three miles. This eighty-foot-thick layer of rare, flaky, blue-gray shale and limestone was once the floor of an ancient sea. ("Lias" comes from the way local stonecutters pronounced the word "layers.")

Henry de la Beche illustrated the fossils that he found with Mary Anning and William Buckland. *(Courtesy of the National Museum of Wales)*

Mary also studied modern creatures, dissecting dead squid and cuttlefish. The more she studied, the better idea she had how the muscles and bones of ancient sea creatures might have been connected and moved, what they ate, and how they lived.

Dissecting animals was a strange way for a girl to spend her free time, and this was hardly the only peculiar thing Mary did. In 1815, the sailing ship *Alexander* sunk not far from Lyme on its way home from India, and the body of a beautiful lady washed ashore. The death of this unidentified woman touched Mary. She pulled seaweed from the lady's hair and visited her corpse daily at St. Michael's Church to cover it with fresh flowers every day until it was finally claimed. Sixteen years later, people in Lyme were still talking about Mary's visits to the corpse.

After the French Emperor Napoleon was defeated at Waterloo in 1815, the agrarian economy of Dorset sunk deeper into depression. Now that the war was over and it was not necessary to pay for the large army and navy, the government abolished the income tax that had been levied on the rich and increased the indirect taxation on food and other necessities that was primarily paid by the lower classes. Before the end of the decade, half a million people were unemployed and the tensions between the working class, middle class, and the aristocracy led to violence. Rioting broke out at Spa Fields in North London in 1816, when a mob armed themselves with weapons looted from a gunsmith's shop. The following year several hundred angry and desperate citizens marched on Nottingham in what became known as the Pentrich Rising.

In March of 1816, five thousand weavers met at St. Peter's Field in Manchester. They began walking peacefully to London to present a petition to the prince regent, but armed cavalry charged at them and scattered them. Two years later, sixty thousand people gathered again at St. Peter's Field. They demanded the right to vote and protested the fact that the growing industrial city of Manchester did not have a single representative in Parliament. This time the troops opened fire and shot unarmed citizens. Eleven people were killed, four hundred were wounded, including more than one hundred women. There were widespread outcries against the murders, but the government imposed censorship and banned public assemblies. In Lyme Regis, Mary Anning and her family continued to face poverty, despite her success in finding fossils.

Chapter Three

The Fish Lizard

It took years for scholars to decide what to call the creature Joseph and Mary had found in the cliff in 1811. Charles Konig, who bought it from Bullock's and placed it in the British Museum, thought it looked half like a fish and half like a lizard. Putting together the Greek words for "fish" and "lizard," he suggested the name *Ichthyosaurus*, meaning "the fish lizard." The name stuck, even though it turned out to be neither a fish nor a lizard. Much later, modern scientists would agree that ichthyosaurs began to swim in the ancient seas during the early Mesozoic era, or the Age of Reptiles, which occured between 65 and 225 million years ago.

When Sir Everard Home, the surgeon to the king, described to scientists the ichthyosaur, he thanked those who bought the fossil without mentioning that Mary and Joseph had found it. He then wrongly identified it as a fish and incorrectly praised Bullock's London Museum for cleaning the fossil.

It took several more years to prove that the ichthyosaur was, in fact, a swimming reptile, but what was clear by 1818 was that Mary and Joseph had found the first nearly complete skeleton of a fossilized reptile to be accurately described by scholars.

The Annings continued searching for fossils, although they were going broke. A retired lieutenant colonel, Thomas James Birch, met the Annings while spending his army pension buying fossils. Birch grew fond of the Annings—and particularly fond of Mary, then in her teens. Concerned by their poverty, Birch wrote on March 5, 1820, to Gideon Mantell, a doctor and geologist, "I am going to sell my collection for the benefit of the poor woman and her son and daughter at Lyme who have in truth found almost all the fine things which have been submitted to scientific investigation: when I went to Charmouth and Lyme last summer I found these people in considerable difficulty—on the [verge] of selling their furniture to pay their rent."

The fossil auction at Bullock's London Museum drew a record crowd from all over Europe. Specimens were purchased both by English buyers and by Georges Cuvier from Paris. The sale gave the Anning's fossil business some much-needed publicity. Birch's collection sold for over four hundred pounds which he gave to Molly Anning. For a while the family had no worries about money.

Mary may have used some of Birch's gift to buy a hat. In 1821, a traveling artist drew her silhouette, showing Mary as a tall, slender young woman with a pretty profile, wearing a high bonnet. She also allowed herself another luxury. As George Roberts was writing the first history of Lyme Regis and looking for subscribers to pay for printing it, Mary Anning was among the first to order a copy.

Apart from Colonel Birch's gift, selling fossils remained the Anning family's only income. Molly ran the little shop in their cellar and Joseph looked for fossils when he could. During the 1820s, Mary gradually took over most of the business.

In May 1821, the Annings dug out of the cliffs a beautiful little ichthyosaur only five feet long. That same month Mary found another ichthyosaur skeleton nearly twenty feet long. Scholars were beginning to notice differences between the specimens. They could not all belong to the same "species," a group of very similar plants or animals. Buckland and de la Beche's friend William Conybeare, a clergyman in Wales and a member of the newly formed Geological Society of London, said that ichthyosaur teeth alone indicated there must be at least four different species of these creatures.

Mary had an extraordinary ability to spot a valuable fossil in a tiny bit of stone poking out of a cliff. One visitor claimed she could look at fifty clumps of rock protruding from a cliff and "pick without hesitation the one which, being cleft with a dexterous blow, should show a perfect fish embedded in what was once soft clay."

Mary was equally skilled at unearthing specimens without damaging them. Finding a fossil was only the beginning of the job. Today fossil hunters use jackhammers and other power tools to unearth and clean fossils. In the 1800s, they had only picks and shovels, hammers and chisels, small brooms, and their own strength. Some of the most common fossils in Lyme, such as an Ammon's horn, which became known as ammonites, were too heavy and encased in stone too hard for most collectors to remove.

It could take weeks or even months to dig out a fossil, wrap it for moving, clean it, set it in plaster, and mount it in a wooden display case. Some collectors hastily reassembled skeletons, but Mary prepared fossils for display as carefully as she excavated them. As difficult as this work was, the Annings inspired countless others to hunt for fossils. Their finds touched off the very first fossil frenzy, decades before there was such a word as "dinosaur."

Despite the excitement their discoveries generated, the Annings continued to struggle. In September 1821, Molly begged Charles Konig at the British Museum to pay for an ichthyosaur he had bought. She wrote, "As I am a widow woman and my chief dependence for supporting my family being by the sale of fossils, I hope you will not be offended by my wishing to receive the money for the last fossil, as I assure you, Sir, I stand much in need of it."

The Annings' sales to prestigious institutions did not gain them much recognition. Neither museum displays nor scientific articles ever mentioned them. The honor went to the men who purchased specimens, not to those who found them.

In her early twenties, Mary had to borrow books and magazines to see what the "learned gentlemen" said about her finds. There were no libraries or museums with these materials anywhere near Lyme, and she could not afford to buy scientific journals or books or pay for memberships in the scientific societies that published the journals. A subscription to the Geological Society of London cost almost two month's income for a poor Dorset family.

Mary's poverty was only one strike against her. A female contemporary wrote in 1823 that she regretted "having been born a woman, and deprived of the life and position which, as a man, I might have had in this world!" Marriage was the goal of every "respectable" Englishwoman. Growing old as a "spinster" was a terrible fate. In the early 1820s, Mary fell in love with an unknown man whose wealth she hoped would lift her out of her "low situation in life." Whoever the man was, Mary "suddenly saw those hopes blasted." She later confided to a friend that it took her eight years to recover from this heartbreak. Combing the shore, climbing cliffs, digging out fossils, and preparing them kept her busy.

Her father's death, her family's poverty, and her lost love could have made Mary bitter. Instead she developed a compassion for others who suffered. A local fisherman who sometimes helped Mary dig out her finds remembered her fondly:

> Sometimes, you know, after a storm—Miss Anning mostly went out after storms—a keg of whiskey might be washed up on the beach—they were smuggling times you understand, and rare times they was. Well, she'd say nothing to the customs-house officers . . . but she'd just cover it up all over with seaweed, and when she comes back from fossiling she'd say to one of us, quite natural like, 'There be a great heap of sea-weed down there, there be' in such or such a place, she'd say. We'd know what she meant. She was wonderful good to the poor, she was.

By 1823, Mary began to receive a little recognition. She sold an exceptionally well-preserved ichthyosaur skeleton to a group of geologists who gave it to the new Bristol Institution for the Advancement of Science, the first such donation the museum received. The opening lecture at the institution called this specimen "the most valuable in the kingdom," but never mentioned who found it. This prompted the local geologist and fossil collector George Cumberland to write to a Bristol newspaper praising "the persevering industry of a young female fossilist, of the name of Hanning . . ." (He spelled her name the way local people pronounced it.) Cumberland told how she snatched from the cliffs "relics of a former world . . . at the continual risk of being crushed by the suspended fragments they leave behind—to her exertions we owe nearly all the fine specimens of Ichthyosauri of the great collections . . ."

Chapter Four

The Monster on the Beach

In December 1823, when Mary Anning was twenty-four years old, she stunned the scientific world with a new discovery. A little west of Black Ven, she unearthed a bizarre skeleton. The creature measured nine feet long and six feet wide but had a head less than five inches long. Hearing of Mary's new discovery, Conybeare was so excited that he could barely finish his Sunday sermon before rushing to see it. Within days he described to the Geological Society of London "the magnificent specimen recently discovered at Lyme."

When de la Beche and Conybeare had examined some fossil fragments in 1821, including one specimen that was missing only its head, they had concluded that another large ancient sea creature besides ichthyosaur must have once lived. Conybeare and de la Beche called it *Plesiosaurus,* which means "almost like a lizard." Its body was shaped like a turtle's, but without a shell. It had a thin neck as long as the rest of its body, but a very short tail. Most startling, its neck had forty bones, far more than any modern reptile or bird. It had four paddles, and these, too, were extraordinary. Few animals have more than five bones in any finger or toe, but *Plesiosaurus* had ten in each digit of its flippers. Buckland said it looked "like a snake pulled through a turtle."

Georges Cuvier, "the father of paleontology," refused to believe such a tiny head could have been attached to so large a body and, furthermore, that a creature could have such a long neck. He initially told Buckland he thought the fossil was a fake, but after studying the detailed drawings made by Mary Moorland (whom Buckland soon married), and hearing descriptions of how Mary Anning had carefully removed the skeleton intact, he declared it, "the most amazing creature that was ever discovered." The duke of Buckingham paid £110 for the fossil, as one newspaper reported, "a price no fossil ever, perhaps, produced before."

Once again, Mary received little praise, at least not immediately. Conybeare began his speech to the Geological Society by thanking the duke for letting scientists see the fossil. He congratulated "the scientific public," rather than Mary, for the discovery. The *Bristol Mercury* reported the find and the duke's purchase but said not a word about Mary.

A few weeks later, though, when Conybeare described this creature to the Bristol Philosophical and Literary Society, he gave Mary credit for its discovery. Felix Farley's *Bristol Journal* did the same a month later. News spread quickly among geologists and collectors that a young woman in Lyme had made this re-markable discovery. The *New Monthly Magazine* called Mary Anning "the well-known fossilist, whose labours lately have

The "father of paleontology," Georges Cuvier, suspected that Mary's *Plesiosaurus* was a fake. *(Courtesy of the Natural History Museum, London)*

enriched the British Museum, as well as the private collections of many geologists . . ."

Finding *Plesiosaurus*—and overcoming charges of fraud from one of the world's greatest scientists—firmly established Mary's reputation as a first-rate fossil hunter. Her discovery, however, also confounded scientists, since it was so unlike any modern creature. Ichthyosaurs, they thought, resembled crocodiles. They had no idea how wrong they were. With its long neck and large body, *Plesiosaurus* was like nothing ever seen before.

Later, modern scientists would find that plesiosaurs lived during the middle of the Mesozoic era.

In the mid-1820s, however, most people, and nearly all geologists, believed ancient animals had been much like modern ones. Some thought that ancient animals were not extinct, but still lived in undiscovered areas of the world. The predominant theory of life on Earth explained that animals were part of a "Great Chain of Being" that began with the simplest organisms and ended with humans, the most complex—and perfect. Trying to marry his religious beliefs with this new evidence of ancient life, William Buckland proposed that the fossils he and others found were created during the Great Flood recorded in Genesis. In opposition, Georges Cuvier argued that Earth had been subjected to many catastrophes that caused the extinction—and creation—of thousands of species. Jean-Baptiste Lamark proposed the most radical theory, transmutation. He believed that an animal could develop into a new and more complicated species over the course of one lifetime by having to adapt to a new environment.

The name "fish lizard" placed the Annings' first discovery between fish and lizards in the Great Chain of Being. The name "almost like a lizard" incorrectly suggested that *Plesiosaurus* was more like modern reptiles. Conybeare and de la Beche called it "a link between ichthyosaurs and crocodiles"—which it was not. Charles Lyell wrote to Gideon Mantell about *Plesio-*

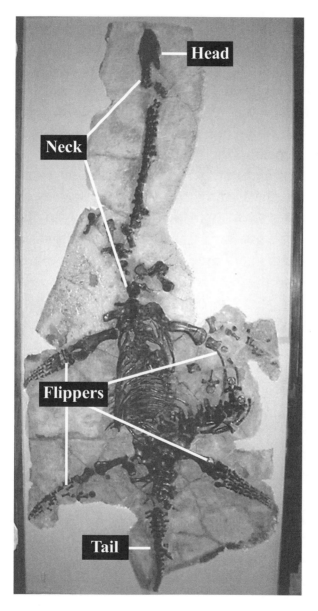

Plesiosaurus dolichodeirus had a large body with flippers, a long neck, and a tiny head. Mary found the first entact *Plesiosaurus* skeleton, pictured here. *(© The Natural History Museum, London)*

saurus, "What a leap have we here, and how many links in the chain will geology have to supply."

Mary, undeterred by the scientific wrangling, continued digging up fossils that fit no pattern previously imagined. She often recognized the uniqueness of her finds before anyone else. She found ichthyoduralites, or "fish-spears," the fin-bones that protected the primitive shark *Hybodus* (which William and Henry wrote about together). She discovered four species of ammonites and studied both modern and ancient fish, even though she lived far from any museum, university, or big library. She wrote to the British Museum in 1824 asking for a full list of its collection and taught herself French so that she could read Cuvier in his original language.

By 1824, Mary Anning was confident enough of her own opinions to argue with William Buckland, who was now Oxford University's first professor of geology. She had just uncovered a fine specimen of *Hybodus*, the first one found with hooked teeth. Thomas Allan, a banker and amateur geologist from Edinburgh, bought the dorsal fin. He wrote in his journal, "Mary Anning's knowledge of the subject is quite surprising—she is perfectly acquainted with the anatomy of her subjects, and her account of her disputes with Buckland, whose anatomical science she holds in contempt, was quite amusing."

Science is based on testing theories and arguing

over whether new evidence fits old ideas. Allan knew that scientists who respected each other still often had forceful debates. What was inconceivable to Allan was that a poor, uneducated young woman would argue with an esteemed scientist. Mary's quarrel with William was a debate between friends. Her letters to him were always affectionate, and he continued to visit her, often bringing his children to search the shore with her.

Mary's willingness to argue with scientists was all the more remarkable when compared to how much ignorance and superstition there was among her neighbors. When a blight, or grub worm, attacked blackberry leaves near Lyme Regis in the mid-1820s, many local people believed the destruction was caused by a flying serpent. A rumor spread that this serpent also killed those under thirty years old, and many rushed to buy magic charms to protect themselves.

Scholars and collectors flocked to Lyme Regis for information as well as fossils. In July 1824, the French geologist Louis-Constant Prévost and the young Scottish lawyer and amateur geologist Charles Lyell saw "a magnificent specimen of an Ichthyosaurus discovered three weeks earlier by the celebrated Mary Anning" and "witnessed the discovery of a superb skeleton" of the common ichthyosaur, only two feet long.

That same month, Mary sold fossils, including an ichthyosaur, to Philip Egerton and William Willoughby, and led the two men along the cliffs. Egerton and

Willoughby had become friends while studying geology at Oxford with Buckland and Conybeare. Egerton spent a vacation with Conybeare at Lyme, where he began collecting fossils with Mary's guidance. She later unearthed—and recognized—the teeth that were missing from the ichthyosaur she had sold Egerton and sent these to him. Mary also shared with Egerton and Willoughby her knowledge of anatomy and suggested how she thought the bones of ancient fish were connected. Her finds and insights helped the men become pioneers in the study of fossil fish.

Lady Harriet Silvester arrived on September 17, 1824, to meet "the famous fossilist reckoned the best in England. By reading and application she has arrived to that degree of knowledge as to be in the habit of writing and talking with professors and other clever men on the subject, and they acknowledge that she understands more of the science than anyone else in this kingdom."

The duke of Buckingham bought two ichthyosaurs that Mary found in 1824. One was nearly thirty feet long. Each new skeleton helped scientists see how bones were connected and how ichthyosaurs had moved. One she found in July, for example, (later named "slender fin and slender jaw") proved that Conybeare was right: There were several distinct species of ichthyosaurs.

Chapter Five

The Old Fossil Depot

On November 22, 1824, when the weather was usually sunny in the corner of Dorset, the worst storm in centuries struck Lyme. Hurricane-strength winds slammed into the coast. Twenty-three feet high waves crashed against the cliffs. Sailing ships were driven aground and homes along the docks were smashed to rubble. The Cobb, which normally protected the village, was severely damaged, as were four dozen houses and businesses. The Annings' first floor washed away. One newspaper reported without exaggeration, "A tempest teeming with more frightful terrors is scarcely within the memory of man . . ." Mary wrote to a friend: "It is quite a miracle that the inhabitants saved their lives. Every bit of the walk from the [Assembly] rooms to the Cobb is gone and all the back parts of the houses on the shore side of town . . . All the coal cellars and coals being gone and the Cobb so shattered that no vessel will be safe here, we are obliged to sit without fires this winter, a cold prospect you will allow."

With Mary's growing fame, she could afford to look for a new home. In 1826, she was able to buy a house on upper Broad Street, just down from the Philpots. In Lyme Regis, moving uphill meant climbing in status. Mary and her mother lived in the back of the house and turned the front into a shop. A small white sign read, "Anning's Fossil Depot."

Mary's renown also led to new friendships. In 1825, Charlotte Murchison persuaded her husband, Roderick, to give up fox hunting and take up fossil hunting. They took their first geological expedition along the coast of southwest England. After a short visit in Lyme, Roderick left his wife there for a while "to enjoy the sea-air . . . and to become a good practical fossilist, by working with the celebrated Mary Anning of that place, and trudging with her, [overshoes] on their feet, along the shore."

With Mary as her teacher, Charlotte Murchison spent years searching for fossils, both with her husband and on her own. She identified and labeled specimens, sketched fossils and the rock formations where they were located, and became Mary's life-long friend.

Helena Emma Waring, who often visited the Depot as a child, recalled Mary and her store:

Our pocket money was freely spent on the little

Ammonites which [Miss Anning] washed and burnished till they shone like metal, and on stones which took our childish fancy. She would serve us with the sweetest temper . . . She was thin and had a high forehead and large eyes which seemed to me to have a kindly consideration for her little customers. There was 'Mrs. Anning the Fossilwoman's mother,' too, a very old lady in a mob cap and large white apron, who sometimes came with feeble steps into the shop to help us with our selection.

Top of the Hill was an interesting neighborhood. Cows pastured nearby and there was an annual cattle fair. A few blocks downhill was the Lion Hotel, where many customers stayed, and the historian George Roberts lived across the street from the Fossil Depot. Around the corner was Sherborne Lane, a steep, one-thousand-year-old cobblestone-paved walkway winding from Broad Street down to the River Lym. It was a good place for a mother and daughter who were selling fossils—and making history themselves—to live.

Owning their own house rather than paying rent helped the Annings through hard times. Sixty English banks went bankrupt in 1825 and 1826, creating nationwide panic. Many rich men who had bought fossils, including the duke of Buckingham, now had no money to spend. (The duke was so embarrassed by his debts that he left England for two years.) Few public museums bought specimens and the price of fossils fell.

Though Mary was better known, few customers could make the trek to Lyme Regis. Until the railroad came to a nearby town years later, the stagecoach remained the fastest means of travel, and coach trips were slower than they had been in Roman times because the English roads were poorer.

Mary had trouble selling even her best specimens. It took a year to sell a "very superb skeleton" of an ichthyosaur, and then she received less than half of her asking price. Because of the difficulty of finding buyers, Mary started selling fossils through merchants in London and other cities. These "brokers" or "agents" usually kept one-tenth of what the purchaser paid as their commission for arranging a sale. In 1826, it was so hard for Mary to sell fossils that she had to pay commissions of twenty percent, and sometimes as high as thirty percent. By the middle of the year, she was deeply discouraged. She told one broker, George Brettingham Sowerby, that she planned to give up fossil hunting.

Instead of quitting, Mary Anning improved her marketing skills. Because she knew there were more buyers for smaller specimens than large ones, she concentrated on excavating small fossils. She also learned from her mother that it was important to write to customers. She had little chance to attend school and had never learned to write well. The Dorset dialect she spoke, along with her thick West Country accent, made her uncomfortable talking to educated gentleman from London—let alone writing to them. Mustering her courage, though, she fired off one letter after another to potential buyers.

In the letters Mary freely shared what she knew. She alerted Henry de la Beche in 1826 to the existence of a submerged petrified forest off Charmouth, visible only at low tide, which she had spotted near the mouth of the river Char. She was quick to learn the Latin and Greek names of newly discovered species. Sometimes she enclosed sketches of the fossils she found and shared her ideas about them. She was, as Mrs. Stock said, "a being of imagination—she has so many ideas and such power of communicating them." She told Adam Sedgewick, who bought fossils for Cambridge University, that one ichthyosaur was "as perfect as if just skinned" and that another fossil was "worthy a place in a museum." This was more than a sales pitch. Mary had enthusiasm for her work and conveyed her excitement to others.

Through her letters Mary made friends with important scholars and their wives. The wives often were more help than their husbands, some being scientists themselves, although seldom recognized as such. She thanked Charlotte Murchison, for example, (who was a first-rate geologist, though Roderick received the glory) for mentioning to several collectors a plesiosaur Mary had found. Soon Sedgewick wanted to buy it.

The "gentlemen of science" were often disdainful

toward those who earned their living selling specimens. Anning also had to overcome prejudice against women, those of the lower class, and those with little education. Scholars nonetheless began to praise her finds, which fetched higher prices for them. Geologists soon began accepting her discoveries without question, a sign of respect they seldom gave other fossil dealers.

Mary often forced scholars to look at familiar objects in new ways. Long, torpedo-shaped fossils called belemnites, "arrowheads," or "thunderstones" were common along the Dorset coast. These hard shells were similar to the "cuttlebone" of cuttlefish, which is found today in parakeet cages. Sometime before 1826, Mary carefully cut open one of these, just the way her father taught her to saw ammonites in half. Inside she found a tiny chamber and what looked like dried ink.

Elizabeth Philpot scraped out this purple powder, ground it up, and added enough water to turn it into a dark liquid. Mary and Elizabeth then dipped brushes into this sepia-colored liquid and drew with it. Mary Buckland, William's wife and geology partner, was startled to receive their sketch of one ancient creature, an ichthyosaur, drawn with ink made from another. This discovery was great for Lyme's tourism business. Local artists began churning out drawings of prehistoric beasts sketched in belemnite ink.

Mary sold William Buckland a number of belemnites with ink-sacks, some nearly a foot long. He knew that thunderstones were the hard shells inside creatures that resembled modern cuttlefish. Studying this prehistoric ink, Buckland saw what Mary Anning and Elizabeth Philpot recognized long before any scientist: These ancient creatures hid from predators by squirting a cloud of ink into the water around them.

The Anning's Fossil Depot drew visitors from near and far. In 1827, the pioneer American geologist George Featherstonhaugh came to collect specimens and meet Mary Anning. Enchanted by this "very clever, funny Creature," he bought many fossils for the new Lyceum of Natural History in New York City. After Featherstonhaugh returned to London, Buckland made additional purchases for him, including a mass of five-sided stone lilies (pentacrinites) she had not yet excavated. Featherstonhaugh's visit led to more orders from the United States.

Even some of Mary's smallest finds revealed secrets of the past. People around Lyme often picked up strange little lumps on the beach they called "Bezoar stones," because they were shaped liked the gallstones of Bezoar goats. Most were dark gray, two to four inches long, an inch or two in diameter. They were twisted in tight spirals, some had black spots.

William Buckland thought they were formed around bits of clay and were not very old. Mary Anning, though, kept finding these Bezoars inside or near the skeletons of ichthyosaurs. As early as 1824, she recognized they were fossilized dung. In 1828, she studied these Bezoars with Buckland. Together, they proved that these were fossilized clumps of undigested food that stayed in an animal's intestines when it died or was expelled in death. Buckland named them coprolites, or "dung stones."

Knowing how coprolites were formed made it possible to see the shape of an ichthyosaur's stomach. The tight spirals of coprolites suggested spiral-shaped intestines, with masses of tiny suction cups indicating that they ate octopus-like squids. The scales, bones, and teeth showed that they also ate *Dapedius*, a primitive fish with thick, shiny scales. The coprolites' black spots, Buckland reasoned, must have been formed from eating belemnites with their sacs of ink.

Ichthyosaurs, in other words, could eat fish and reptiles protected by sharp teeth, thick scales, hard shells, and camouflage. They must have been formidable hunters. Most surprising of all, backbones of smaller ichthyosaurs were found in their fossilized feces. These creatures were cannibals.

Mary enjoyed a good argument among educated gentlemen. She wrote to Charlotte Murchison, "I do enjoy an opposition among the big-wigs." Coprolites shook up all who studied prehistoric life. Until this time, scientists had little idea how living creatures became fossils. If something as soft as undigested food could be preserved, then some fossils must have been created by "the sudden entombment" of animals, before even the softest tissue could decay.

A fossil, they learned, was not a creature which had turned into stone. Instead a fossil is part of an ancient animal or plant (or its outline) which is preserved from decay. A creature or plant would be covered by mud. The air would gradually be squeezed out of the body, thus keeping the creature from rotting. It might take thousands of years for the mud to become compressed into layers of mudstone, but the mud itself could quite quickly seal up bones or tree trunks, skin or fur, bark or leaves. In this way, Mary's discovery of coprolites led scientists to a new understanding of how fossils were formed.

Chapter Six

The Flying Dragon and the Winged Fish

For some time Mary concentrated on smaller, easy-to-sell, fossils. She and Elizabeth Philpot spent a great deal of time in the grip of what Mary called "the green Sand Mania," looking for fossil sea shells in the higher (and newer) layer of the cliffs called Greensand. They invested countless hours "beating bits of green Sand to pieces to find shells . . ." Finding little of note and bored with sea shells, Mary gave up on Greensand and returned to the lias. There, in November 1828, she made another small but significant find, "an unrivaled specimen of Dapedium politum, an antidiluvian fish," called "pavement fish" because its rectangular, jet-black scales resembled paving stones.

In December 1828, when Mary was twenty-nine years old, she uncovered the first flying reptile ever found in England. After writing ten volumes about prehistoric animals, Georges Cuvier said about pterosaurs: "Of all the ancient beings which have been discovered, these were undeniably the most extraordinary, and those which, if one could see them alive, would seem the most unlike anything living today."

William Buckland told the Geological Society of London that Mary had found an "unknown species . . . a monster resembling nothing that has ever been seen or heard-of upon earth . . ." The fossil was the size of a raven, three and a half feet long (a little over a meter). Its skin-covered wings stretched from its body out to the long fingers of its forearms. Like modern birds, it had light, hollow bones. Its enormous head had massive but lightweight jaws with four big, sharp teeth, spaced far apart, at the front of each jaw, good for cracking the tough skins of small land animals or the inner shells of squids. Its back teeth were smaller but equally sharp. Eventually scholars named it *Dimorphodon*, which means "two forms of teeth." Modern scientists place *Dimorphodon* during the middle of the Mesozoic era, making it a contemporary of plesiosaurs and ichthyosaurs.

Mary's *Dimorphodon* was the first pterosaur, or "winged lizard" found outside of Germany. Before her discovery, only two kinds of flying reptiles had ever been found. With the discovery of *Ichthyosaurus*, *Plesiosaurus*, and *Dimorphodon*, Mary Anning's fossil business was now famous among British scientists and was becoming well known elsewhere. In February 1829,

Charles Lyell announced her discovery to Louis-Constant Prévost's students in Paris.

Early in 1829, Mary found her second complete plesiosaur skeleton, larger and better preserved than the first. Scientists dubbed it "the long-necked plesiosaur." The *Salisbury & Winchester Journal* announced: "The lovers of science will be delighted . . . with the light thrown upon the antediluvian [prehistoric] world by a gratifying discovery made at Lyme by Miss Mary Anning, of another specimen of the Plesiosaurus . . . unequaled in any country." A museum in Philadelphia planned to purchase Mary's new plesiosaur. William Buckland was furious when he learned England might lose it, and he demanded that the British Museum buy it. The museum rescued it for England, paying £105.

As Mary turned thirty, she may have needed some extra praise. Unmarried English women were thought to be "old maids" by thirty, and were often treated with pity or contempt. To make matters worse, Joseph, her only surviving brother, married in 1829—and there was tension between Mary and her new sister-in-law.

It pleased Mary when the *Edinburgh Philosophical Journal* included her in 1829 (along with the Philpot sisters, Charlotte Murchison, the duke of Buckingham, George Cumberland, Gideon Mantell, and Adam Sedgewick) in the first ever published list of geological collections in Great Britain. That same year, Cumberland wrote in the Royal Institution of Great Britain's *Quar-*

terly Journal of Literature, Science, and the Arts about her "industry and skill" in finding "nearly all the fine specimens" of ichthyosaurs that had been found.

Mary longed to travel. She had told Charlotte Murchison earlier that she was very sorry that she could not accept an invitation to visit her in London because "I have never been out of the smoke of Lyme." In July 1829, she finally visited the Murchisons in London. She went to the Geological Society, the British Museum—"with which I was much delighted"—the British Library, Sowerby's "museum," a diorama of Rome, and another of the Vatican. She attended a worship service at an Anglican church and heard the Bohemian Brethren (a Czech religious group) sing.

At the end of the year, Charles Lyell (who would become the most important geologist of the century) wrote for her help in determining whether the sea was encroaching on the Dorset coast.

In December 1829, Mary made another important find. The *Salisbury & Winchester Journal* reported: "It is about a foot and a half long; has two immense sockets for the eyes, a long snout, a series of finest vertebrae that have ever been seen in so small a creature, claws, fins like wings . . ." Some scholars said the fossil was a reptile, others thought it must be a bird. When they finally agreed four years later that it was a fish, they continued to argue about its characteristics. They named it *Squaloraja*, which mistakenly emphasized its supposed similarity to modern sharks and rays. Mary dissected a modern ray and saw that her fossil was a fish of an entirely new species. She tried to entice Adam Sedgewick at Cambridge into buying it, saying, "It is quite unique, analogous to nothing."

Mary was finally receiving credit for her contributions to science. In 1830, the geologist Edward Pidgeon wrote in his book about fossils, "It is to her almost exclusively that our scientific countrymen . . . owe the materials on which their labors and fame are grounded." However, 1830 turned out to be a bad year for selling fossils. Once again there were few rich men with money to spend on luxuries, and with revolution in Belgium and France and more bread riots at home, those who had money were reluctant to spend it. Mary found another ichthyosaur that year, which the Bristol Institution bought, but the prices of these specimens continued to fall. Soon the Bristol Institution was facing a financial crisis as local banks went bankrupt. Many scientists were interested in Squaloraja, but it took Mary a year and a half to find a buyer. She even had trouble selling Dimorphodon, despite its rarity. In the end, William Buckland had to purchase it himself to save it for the cash-strapped British Museum.

Dorset was particularly hard hit by the economic depression, facing the sort of poverty during the 1820s and 1830s that is prevalent today in third-world nations. Few people had savings to cushion them in hard

Henry de la Beche's painting featured every ancient creature imagined to have once lived in Dorset. *(Courtesy of the National Museum of Wales)*

times, and the only governmental safety net was meager "poor relief." Despite the county's reliance on farming, famine stalked the land. Once again Mary's family became impoverished.

Henry de la Beche drew a picture to raise money for the Annings. Called *An Earlier Dorset*, it showed how he imagined prehistoric life had been. His was the first real attempt to document what scholars knew about prehistoric life, and it had an enormous impact on how everyone after him illustrated scientific books. Henry packed into *An Earlier Dorset* all the creatures the Annings had found along with some reptile dung. One ichthyosaur attacked a plesiosaur while others devoured a *Dapedius* fish and a belemnite. A *Dimorphodon* flew over these battles while a *Hybodus* swam below them. An ammonite swam along the surface of the sea. Below it, a flower-shaped crinoid or "sea lily" that Mary had found was anchored to the ocean floor by its five-sided stem.

George Scharf, one of the finest scientific illustrators in London, made Henry's drawing into lithographic prints. Copies were sold for today's equivalent of about one hundred dollars. Buckland used an enlarged print in his geology lectures at Oxford, further increasing sales.

In December 1830, Mary made her fifth great discovery, the skeleton of a new species, eventually named "the plesiosaur with the bigger head." It had even more neck bones than the specimen she found in 1823. William Willoughby bought it for £210, a remarkable price, equal to at least ten thousand dollars today. It would be almost a decade before the skeleton was accurately described. Even then, Richard Owen thanked Willoughby for letting him examine it but failed to mention who had found it. He referred to it as "Hawkins' plesiosaur" because Thomas Hawkins, an eccentric collector, had proposed a name for it.

Meanwhile, William Buckland was obsessed with finding more dung stones. His friend Henry de la Beche appreciated their importance but could not resist poking fun at Buckland's "coprolitic vision." Around 1829, he produced a lithograph that showed Buckland, the reverend professor of mineralogy and geology in the University of Oxford, dressed in academic gown and hat, standing at the entrance of a cave shaped like a cathedral, raising a geological hammer as if leading a choir. Every member of the choir and congregation are caught in the act of defecating—crocodiles, ichthyosaurs, pterosaurs, and even Buckland himself.

Buckland had told the Geological Society that he hoped coprolites might someday be found from plesiosaurs, which themselves are more rare than ichthyosaurs. He was excited when Mary wrote to him: "last week I discovered a young Plesiosaurus about half the size of the one the Duke [of Buckingham] had. It is without exception the most beautiful fossil I have ever seen . . . the head is twice as large in proportion as those I have hitherto found. The neck has a most graceful curve, and what makes it still more interesting is that resting on the bones of the pelvis is its Coprolite."

Buckland found the discovery of these tiny coprolites profoundly moving. No stranger to coarse humor, he made a tabletop inlaid with polished slices of preserved reptile dung. Buckland wrote:

> When we see the body of an ichthyosaurus still containing the food it had eaten just before its death, and its ribs still surrounding the remains of feeding

that were swallowed ten thousand, or more than ten thousand times ten thousand years ago, all these vast intervals seem annihilated, come together, disappear, and we are almost brought into as immediate contact with events of immeasurably distant periods as with the affairs of yesterday.

Coprolites were a window to the past. One of William Buckland's students at Oxford, John Shute Duncan (1769-1844), who later became the head of the Ashmolean Museum at Oxford, wrote a poem about the creature:

Approach, approach, ingenuous youth,
And learn this fundamental truth
The noble science of Geology
is bottomed firmly on Coprology.

By 1831, a new friend noted that Mary had been "noticed by all the cleverest men in England, who have her to stay in their houses [and] correspond with her on Geology." Despite the attention, Mary continued to work hard. She ended a letter to J.S. Miller at the Bristol Institution by saying that she must hurry back to the shore: "The tide warns me I must leave of scribilling [sic]."

Her hard work continued to pay off. In the spring of 1831, she found an ichthyosaur, "the best she had ever seen," a specimen William Buckland described as "very perfect."

Mary's work did not end, however, after she had found, excavated, cleaned, prepared, sold, and shipped a fossil. Sometimes she had to pester her customers for payment. She wrote to Adam Sedgewick on May 9, 1831, thanking him for the five pounds he had paid for an ichthyosaur head and the skeleton of another ich-thyosaur—and reminding him that he still owed her another ten shillings. There is no indication that he ever paid up.

It was hard to hound customers about overdue bills without driving them away. Ten shillings may not have seemed like much to a Cambridge professor, but it was critical to the Annings. Mary complained that "those men of learning have sucked her brains, and made a great deal by publishing works of which she furnished the contents, while she derived none of the advantages." As she told a young girl in London, "I beg your pardon for distrusting your friendship. The world has used me so unkindly, I fear it has made me suspicious of every one."

Chapter Seven

The Lioness of Lyme Regis

At the end of the summer, fewer tourists visited Lyme Regis and fewer customers entered the Anning Depot. Fall and winter were the seasons when Mary was most likely to find new specimens. The best time to search was right after a storm had knocked down part of a cliff and exposed fossils. This was also the most dangerous time to be near a collapsing hillside, but this did not stop Mary. She wanted to reach newly uncovered areas before the rock crumbled further.

Mary's young friend Anna Maria Pinney recorded that when she first went fossil hunting with Mary in 1831, they climbed down slopes she thought impossible to descend. The wind was stiff, the ground slippery, and waves beat against the cliff. Once on the beach, they reached a place where Anna Maria thought they could go no further: "Before I knew what to do she caught me with one arm round the waist and carried me for some distance, with the same ease as you would a baby."

Mary impressed Anna Maria with her sharp tongue and great strength. "She glories in being afraid of no one, and in saying everything she pleases," Anna Maria wrote in her diary that night. "She would offend all the world, were she not considered a privileged person . . . She was very good humoured with me, but gossiped and abused almost everyone in Lyme, laughing extremely at the young dandies, saying they were . . . numskulls, not men." A member of the upperclass, Anna Maria suspected that associating with people of higher social class had made Mary dislike people from her own background.

Anna Maria had just arrived in Lyme and may have been put off by the blunt, forthright style of speaking that Lyme's Dissenters valued. Lyme had always been known for its contentious, argumentative inhabitants. If Mary was argumentative and quarrelsome, she was not much different from her neighbors.

Mary's caustic remarks sometimes cost her sales. After a friend of J.S. Miller visited the Anning shop, Mary offered an apology: "I beg ten thousand pardons for my Mother's mistake, it was another Gentleman of the same name whom I had complained." Her explanation—probably false—evidently did not convince Miller. Buckland had thought he would buy *Squaloraja* for the Bristol Institution, but Miller decided not to.

According to her mother, Molly, Mary "had no sweethearts" and sometimes fell into dark moods. Some people even thought that Mary was crazy. Anna Maria's mother refused at first to let her daughter go to the beach alone with "this extraordinary character." The Pinneys gossipped, some sixteen years later, about how Mary had visited the corpse of the drowned woman at St. Michael's Church.

Mary admitted, at least to herself, that she sometimes felt a bit confused. She laughed at herself the same way she laughed at foolish young men. In her journal she wrote: "I can enjoy everything but gloomy silence; conversation I doat [dote] on, whether gay or serious, and if the contents of my head are a little derangee [disturbed], still dear—you must allow there are contents, and time, perhaps aided by my patient perseverance, may work wonders and produce order where hitherto disorder has reigned queen of the empire."

Anna Maria believed Mary's visits to the unclaimed corpse showed "the wild romance of her character . . ." She admitted Mary's behavior was eccentric but believed this was caused by heartbreak: "Her wildness of manner . . . is taken for madness by those who do not understand the agony of blasted hopes . . . it is not surprising that she goes on like a phantom, someone who has suffered as much as Mary has . . ."

Anna Maria quickly noted Mary's rare skill: "She has more than the power of an eagle's eye, when she

This nineteenth-century watercolor sketch shows how Mary often worked alone, searching for fossils in the cliffs around Lyme Regis. *(Courtesy of Roderick Gordon.)*

searches among the sands and rocks." She recognized, too, that Mary took risks others could not imagine because she loved her family intensely:

> She is very kind and good to all her own relations, and what money she gets by collecting fossils goes to them and to any one else that wants [needs] it . . .

> see her with anyone she loves and the lion is tamed:
> she is gentle, attentive, and has a . . . way of express-
> ing her affectionate feelings beyond any person I
> ever met, and at the same time an ardour which
> would esteem all dangers and difficulties trifles . . .
> She will attend the sick [and] poor night and day,
> when they are ill with infectious diseases, and she
> has supported her mother and brother in bitter pov-
> erty, and when she was so ill that she was generally
> brought fainting from the Beach . . . Had she lived in
> the age of chivalry, she might have been a heroine
> with fearless courage, ardour, and peerless truth and
> honour.

A combination of love and pride, wildness, and dis-
regard for others' opinions sealed Mary's commitment
to face landslides and crashing waves. Her career was
enough to convince many people that she was strange.
While other well-off women took an interest in geology, few spent time searching for fossils. Maria Hack
wrote about Mary in 1832, in one of the first geology
books ever published for young people. She devoted a
chapter to Mary's exploits: "It is certainly uncommon
to hear of a lady engaging in such fatiguing, hazardous
pursuit; and I think few would be found willing to
undertake a personal examination of the cliffs, espe-
cially in the depths of winter."

Whatever their concerns about Mary's sanity, Anna
Maria's parents did not hesitate to ask Mary for help

when they needed it. The Pinneys had moved to Dorset in October 1831 so that Anna Maria's twenty-five-year-old brother could run for a seat in Parliament. Before 1832, only men who owned a great deal of land could vote. However, the Reform Act of 1832 extended voting rights to many middle-class Englishmen. William, although quite wealthy—the Pinneys owned slaves and sugar plantations in the West Indies and lived in a great house set on a hill—was a "Reform" candidate running against the landowning aristocrats who had ruled Lyme for decades. William needed the votes of ordinary people to win.

The Pinneys asked Mary and Joseph Anning to campaign for their son, even though Mary could not vote. Mary and Joseph agreed and, with the Anning's support, William Pinney was elected to Parliament repeatedly over the next three decades. Joseph Anning, who worked as an upholsterer, soon became a town leader.

Some people were not at all impressed with the Fossil Shop. Gideon Mantell visited Lyme in 1832 and wrote in his journal: "We sallied out in quest of Mary Anning, the geological Lioness of the place. We found her in a little dirty shop, with hundreds of specimens piled around and in the greatest disorder. She the presiding Deity, a prim, pedantic vinegar looking, thin female, shrewd, and rather satirical in conversation."

Regardless of what others thought of Mary or her shop, she continued to make extraordinary new finds.

In 1832, she unearthed an enormous new ichthyosaur species (dubbed "slender fins and flat teeth") nearly thirty feet long. Even Gideon Mantell had to admit to Mary's talent.

As more ichthyosaurs were found, however, Mary received less money for each one. Twelve years earlier, a small skeleton was auctioned at Bullock's Museum for £152. By 1832, Mary was offering Adam Sedgewick the specimen she called "the best yet discovered" for thirty-five pounds. Soon she would settle for only fifteen pounds for well-preserved skeletons.

Despite falling prices, Mary continued to help other fossil collectors—even those who competed with her for fame and fortune. She led the wealthiest visitors along the shore without charging a fee, although most guides did. Mary often allowed her visitors to keep the fossils that she found on these excursions.

The eccentric collector named Thomas Hawkins visited Lyme for the first time in July 1832. Mary sold him a newly discovered ichthyosaur head and led him to the rest of the reptile, which she could have sold herself. Determined to get the skeleton, Hawkins tore down an entire face of the Church Cliffs to pry it loose. Even though Hawkins was a rival and may have damaged other fossils Mary sought, she helped him box the fossil fragments he unearthed. Along with the surrounding rock, the fossils weighed over a ton.

Mary Anning guided even the youngest fossil hunters. In November, she led a boy Anna Maria called "little Marder," on a long hike beyond Charmouth. They found a sea lily. They looked in a book on all the known crinoids but could not find it cataloged. They had discovered a new species. To their dismay, because no one described this find in a scholarly journal, it went unnoticed. The young boy nevertheless remained excited about his find. James Wood Marder grew up to exhibit and sell fossils in Lyme. He even had a flying reptile named after him.

Unfortunately, scholars frequently overlooked the significance of Mary's finds. Around this time, someone misplaced the coprolite she had carefully preserved with an ichthyosaur skeleton she sold to Adam Sedgewick. Not realizing the importance of this tiny object, geologists at Cambridge lost it—or simply threw it out. The British Museum somehow misplaced all but the head of Mary and Joseph's first ichthyosaur, as well as most of the large one Mary found in 1832.

Many men simply refused to take women seriously in the field of geology. In 1832, debate raged among the members of the Geological Society of London whether or not Charles Lyell should let women attend his geology lectures at King's College. At the time, not a single university in Britain admitted women. Lyell was reluctant to let women listen, even though his fiancée, Mary Elizabeth Horner, was an expert in sea shells and worked with him on nearly all his geological expeditions. Even William Buckland, whose own studies benefited so much from the work of Mary Anning, the Philpot sisters, Lady Mary Cole, Jane Talbot, and his own wife, Mary, (who edited and illustrated his books), opposed admitting women to scholarly meetings. Lyell agreed to admit women only after Roderick Murchison insisted that he must bring his wife and geological partner, Charlotte. Women flocked to Lyell's lectures. Soon, however, the bishop of London barred women from King's College. Finally finding courage, Lyell promptly resigned from the faculty and took his lecture series to the Royal Institution.

In 1832, some men may have wanted to keep women out of geology, but a lithograph Henry de la Beche printed that year indicated their efforts came too late. Titled, "The light of science dispelling the darkness which covered the world," it pictured a determined young woman holding a lamp in one hand and a geological hammer in the other.

Chapter Eight

Praise

In 1832, a worldwide cholera epidemic reached Lyme. Anna Maria Pinney described the epidemic as being "like a sword from the east, has it traversed the earth, in 14 years killing 50 million." Mary Anna Maria, the Philpots, and most of their friends were spared. The following year a fire swept through Lyme, but once again, Mary was not hurt.

Mary's work itself was as dangerous as any disease or disaster. Crumbling cliffs were both the source of her livelihood and a threat to her life. One day in 1833, she went to the beach with her dog Tray, just as she had many times. After the fierce storms that had racked the coast that winter, it seemed to be a good time for fossiling. Her dog, as usual, kept Mary company as she hammered into the cliff. She heard only the cry of birds, Tray's occasional bark, the tapping of her hammer, and the sound of waves. Suddenly she heard an ominous roar above her head. In the blink of an eye, tons of rock fell around her and crushed Tray.

Mary was stunned by the loss of Tray and by her own narrow escape. She told Charlotte Murchison: "The death of my old faithful dog has quite upset me. The Cliff fell upon him and killed him in a moment before my eyes and close to my feet; it was but a moment between me and the same fate."

On December 8, Mary survived another close call. As she headed to the beach before dawn, a large, heavy runaway cart came crashing down the steep road straight toward her. Mary was pinned against a wall. Elizabeth Philpot wrote to Mary Buckland: "Yesterday she had one of her miracle escapes in going to the beach before sunrise and was nearly killed in passing over the bridge by the wheel of a cart which threw her down and crushed her against the wall. Fortunately, the cart was stopped in time to allow her being extricated from her most perilous situation." Mary had long been deeply religious, but after these close calls, Anna Maria noticed, "The Word of God is becoming precious to her . . ."

While Mary was finding comfort in the Bible, her finds made others question it. During the 1830s, people throughout England left their churches. The discovery of ancient fossils undermined faith in the book of Genesis as a literal description of the creation of the world. When Mary was a child, geology seemed to many Dissenters to be a good way to prove the literal interpretation of the Bible. Many religious leaders attacked this new science.

Lyme's Independent Chapel was once one of the most respected Dissenter congregations in Dorset. Wracked with rising doubt and a succession of disappointing pastors, the chapel nearly disbanded during the 1830s. Ebeneezer Smith, the chapel's pastor from 1828 to 1838, left and joined the Church of England. Mary occasionally continued to give money to the troubled Independent Chapel after 1830, but she gradually shifted her interest to St. Michael's, an Anglican church. She also made a small donation toward building the Methodist Chapel on Church Street.

Her friendship with St. Michael's priest, Fred Parry Hodges, and other Anglican clergy such as Buckland and Conybeare, made this transition easier. Many of her friends, including Anna Maria Pinney and Elizabeth Philpot, worshiped regularly at St. Michael's and frequently discussed Hodges's sermons. Becoming an Anglican also indicated a rise in social status. Anna Maria Pinney observed, "Associating with and being courted by those above her, she frankly owns that the society of her own rank is becoming distasteful to her . . ."

Meanwhile, arguments about how to understand fossils and the Bible raged across the nation. Conybeare, who was still a pastor as well as a geologist, defended scientific research. People saw a conflict between geology and the book of Genesis, he claimed, only because they misunderstood the Bible. They might think the earth was formed a few thousand years ago, for example, but the Bible does not say this.

In the early nineteenth century, it was assumed that all of "God's creatures" still existed. People were not familiar with the idea that species became extinct. The earlier discovery, in Europe and America, of mammoth and mastodon fossils had been disturbing. These mammals looked like modern elephants, and scholars correctly guessed that they had lived only a few thousand years ago, but they incorrectly believed that they were still alive somewhere in the vast American West. President of the United States Thomas Jefferson hoped that Lewis and Clark would find living mammoths during their exploration of the Louisiana Purchase.

By the early 1830s, overwhelming evidence showed that strange reptiles had once filled the land, sea, and sky. Some people, such as the Reverend William Kirby, still refused to believe that ichthyosaurs, plesiosaurs, and pterosaurs were extinct. He claimed they would be found in unexplored regions or in caverns deep within the earth. Some geologists, however, found evidence in fossils of a "Divine Creator." William Whewell, a Cambridge professor, said the vast differences between "the plesiosaurs and pterodactyls of the age of the lias" and fossils in more recent rock layers could be explained only by "a distinct manifestation of creative power, transcending the known laws of nature."

On the other hand, the science writer John Murray observed, "among geologists there is a sad preponderance of skepticism." Murray tried to reconcile Mary's finds with the story of Noah and the flood. He insisted, despite her discovery of additional skeletons, that "the plesiosaurus somewhat resembles the crocodile." Murray even rejected the theory that the location of fossils indicated how old they were because this was dangerous to "the truth of revelation." Mary, though, who took Murray fossil hunting herself, kept careful records showing exactly where she made each of her discoveries. As a result, simply visiting the Fossil Depot could upset people.

In 1833, Reverend Henry William Rawlins took his family to Lyme for a seaside holiday. Attracted by a fine display in Mary's window, six-year-old Frank Rawlins stepped inside. Mary wrote labels for the fossils Frank purchased that told him how far below the top of the cliff she had found each one. The boy wondered how different types of creatures could have been laid down in different layers of the ocean floor that later became Lyme's cliffs.

Frank's father believed that God created each species in a single day, with all creation completed in a week. How could this be, his son asked, when different kinds of fossils are found at different heights in the cliff? Frank believed their placement in the cliff meant they lived many years apart. His father refused to discuss Frank's questions, insisting these creatures died in the Great Flood.

Because Mary was deeply religious herself, the inquisitive boy had trouble dismissing what he learned from her. A decade later, as a seventeen-year-old student at Cambridge University, Frank secretly read *Vestiges of the Natural History of Creation*, a book so controversial that its author hid his identity and published anonymously. *Vestiges* argues that life had evolved from simple to complex forms independently. The book was denounced as heresy by some, and heralded as a breakthrough by others. It immediately attracted one hundred thousand readers and received more attention than any other scholarly work ever published. Adam Sedgewick thought the book was so dangerous that he warned Mary Lyell and her sisters not to read it. Frank Rawlins read it and became both a clergyman and a geological collector.

Earlier, Etienne Geoffroy Saint-Hilaire in Paris had laid some groundwork for evolutionary theories, suggesting that modern animals were descended from ichthyosaurs, plesiosaurs, *Dimorphodon*, and other prehistoric creatures. In England, Charles Bell wrote in 1833 that ichthyosaur and plesiosaur paddles showed a progressive development over vast stretches of time as they adapted to their environment—a notion that Charles Darwin later drew upon in shaping his own theory of natural selection.

Around this time, de la Beche painted a watercolor of Mary Anning working on the shore, dressed in sturdy shoes, a heavy coat, and a top hat. Made from felted wool coated with shellac, these hard top hats served as the crash helmets of the time. Early geologists wore them for protection from falling rocks.

The next year, 1834, Mary Anning and Elizabeth Philpot helped the Swiss paleontologist Louis Agassiz study fossil fish. Mary and Elizabeth showed him how to match fossil backbones with teeth found in the same layers of the blue lias. Louis was so grateful that he gave them an honor women seldom received: In his pioneering book surveying all fossil fish yet discovered, he thanked the women for their help. Agassiz also named several species he discovered after Mary, something no English scholar did in her lifetime—though de la Beche, Birch, Buckland, and Henley all had ammonites named after them. What Mary and Elizabeth taught him, he wrote in *Studies of Fossil Fish*, was the key that allowed him to unlock the secrets of prehistoric fish.

Agassiz later moved to the United States, where he became one of the most important scientists of the nineteenth century and helped to popularize the study of natural history. By describing and drawing prehistoric fish, he provided evidence that life on the earth had passed through great "catastrophes." Agassiz suggested that much of the earth was once covered with ice. After Agassiz convinced William Buckland that the

Ice Age had occurred, Buckland persuaded most other English geologists—though it took years to do so.

Many people refused to believe the earth could be old enough for glaciers to have shaped its surface. When Buckland told the British Association for the Advancement of Science that our planet must be at least hundreds of thousands of years old, he was widely condemned by other preachers. Mary, however, continued to uncover evidence that life on earth was older—and stranger—than anyone imagined.

Louis helped Mary with her studies. The ancient shark *Hybodus*, he suggested, might have used its four large, hooked teeth to grab its prey. Then it could have sunk its smaller teeth—a hundred of them—into its prize to chew it up. He also identified the fish scales in Ichthyosaur coprolites. The most common meal of ichthyosaurs, it turned out, was *Pholidophorus*, a small, primitive fish that had hard, enamel-coated scales, found in the lias.

By 1834, George Roberts wrote that Mary Anning was "already well-known in every part of the world where persons read of the discoveries and progress of science." He had not even mentioned her by name in his 1823 *History of Lyme Regis*, but now he praised her "genius for discovering where the Ichthyosauri lie imbedded . . . [her] great judgment in extracting the animals, and infinite skill . . . in their development . . ."

Thomas Hawkins praised her just as lavishly in his

own book the same year: "This lady, devoting herself to Science, explored the frowning and precipitous cliffs there, when the furious spring-tide conspired with the howling tempest to overthrow them, and rescued from the gaping ocean, sometimes at the peril of her life, the few specimens which originated all the fact and ingenious theories."

When members of the Geological Society of London heard a speech in 1835 describing the newly discovered fossil crab *Coleia antiqua* and the ancient "Brittle star" *Ophioderma egertoni*, Mary received credit for finding them. There was no need to explain who she was—they all knew her.

Now thirty-six years old, Mary continued to tackle tough challenges. She told Adam Sedgewick on July 27, 1835, that she had just found a baby ichthyosaur, "the smallest I have yet seen, about 1 foot 9 inches in length." She was digging out a twelve-foot ichthyosaur buried along the shoreline, where "the tide would not allow of our working above one hour in a day." Mary may have worked this hard because she was, once again, desperate for money. She lost her life savings, at least £300, when a man with whom she had invested it died suddenly and left her with almost nothing.

That same year she announced her discovery of a new thirty-foot-long ichthyosaur species. She promptly sold it to Philip Egerton. Once again, a new find led to new knowledge. Sir Philip noticed that some backbones near the creature's head appeared to be fused together. Mary told him that she often saw this in ichthyosaurs. By carefully examining the neck bones in this newest specimen, he realized that their necks could not have moved much but would have been extremely strong. In a speech to the Geological Society of London, Egerton gave Mary credit for the discovery, praising her "zeal and intelligence."

Mary's discovery of ichthyosaur and plesiosaur coprolites led Gideon Mantell and others to look for similar clumps of undigested food in fossil fish. Mary herself discovered coprolites from several species of fish. She also noticed a curious pattern in pentacrinites she found between Lyme and Charmouth. These crinoids were attached to a thin layer of "jet," a coal-like substance, and all were growing in the same direction.

Buckland concluded that the pentacrinites attached themselves to floating chunks of wood. Fossils could be formed slowly, he realized, if a plant or animal died where it did not rot quickly. If enough sea lilies climbed aboard a floating log, he reasoned, they would sink it. On the ocean floor, where there were few scavengers and almost no oxygen to fuel its decay, pentacrinites might be preserved as the mud at the bottom turned into stone. This observation suggested that pentacrinites could crawl like starfish rather than being anchored permanently in place like clams. In 1836, Buckland described this discovery by "Miss Mary Anning, to whom the scientific world is largely indebted."

The next year the popular American writer Samuel Goodrich poured out praise in his book *Wonders of Earth, Sea and Sky* and added, "No one ought to go near Lyme Regis without visiting her collection." The German explorer Ludwig Leichhardt did not miss his chance in 1837. This future explorer of Australia wrote: "We had the pleasure of making the acquaintance of the Princess of palaeontology, Miss Anning. She is a strong, energetic spinster of about 28 years age [perhaps she lied about her age], tanned and masculine of expression."

Chapter Nine

Through the Storm

Mary's fame was now international. After spending three days in Lyme Regis fossil hunting with Mary, poet and philanthropist John Kenyon published a poem celebrating her work:

> Though keenest winds were whistling round,
> Though hottest suns thy cheek were tanning,
> Nor suns nor winds could check or bound
> The duteous toils of Mary Anning.

Mary Anning's labor was rewarded with something more than poetry when the British Association for the Advancement of Science and the British government gave her a "Civil List Pension" of twenty-five pounds a year in appreciation for her work. Between this and the fossils she sold, Mary and her mother were finally able to live comfortably.

Early in 1839, as she neared forty, Mary found "the most perfect jaw of Hybodus." In April of that year she mustered the courage to write to the *Magazine of Natural History* about this prehistoric shark. She wanted to set the record straight that it was she who had discovered, many years earlier, that some fish had both straight and hooked teeth. She reminded scholars of an insight made by Louis Agassiz, which he apparently forgot himself: The hooked teeth "were the teeth by which the fish seized its prey—milling it afterwards . . ."

Three months later she wrote to the journal's editor, arguing that the new *Hybodus* specimen did not belong to a new genus (a broader classification than species) adding, "as I am illiterate, [I] am not able to give a correct opinion." Mary was hardly "illiterate" the way the word is used today, as she could read and write.

However "illiterate," or uneducated, she felt, Mary's discoveries were beginning to have a major impact on the way scientists understood the world. During 1838 and 1839, for example, the young naturalist Charles Darwin wrote about ichthyosaurs and plesiosaurs several times in the notebooks he kept as he formed his theory of evolution, which would become the single most important theory about life on earth.

In September 1839, Richard Owen came to Lyme to see Anning, hoping to flatter her and gain information from her. She led Owen, Buckland, and Conybeare on a geological excursion—one of the few times Owen visited a fossil site himself. Owen was impressed with the way forty-year-old Mary scrambled over the cliffs when the tide threatened to swamp them. A year and a half earlier, Owen had slighted Mary when he described in great detail to the Geological Society of London the remarkable *Plesiosaurus macrocephalus* she had found without mentioning her name.

Early Christmas Day 1839, after months of unusually heavy rainfall, several families of laborers on the Dowlands' Farm west of Lyme were disturbed when the ground beneath their homes began to shift. That evening forty acres of land west of Lyme slid into the sea. This "Great Landslip" opened a chasm 150 feet deep, 400 feet wide, and three-quarters of a mile long. In a single night, eight million tons of rock crashed into the ocean. A month later Mary was the first person to notice that a foot-high ridge was forming on the beach at Whitland and that rock along the cliff was breaking up. Soon another "landslip" left a gash in the earth sixty feet high and a quarter of a mile long.

In addition to running her shop and searching a three-mile stretch of lias, Mary stayed in touch with distant friends. Even when she felt overwhelmed by the mail she had to answer, letters often brought her joy. She wrote on January 13, 1840, about her visit to the landslip, "although I am almost ready to cry at the heap of letters lying before me, still I was truly delighted to hear once more from you." She ended the letter wishing

Charles Darwin's theory of natural selection became the most important theory on how life evolved on earth. *(Courtesy of the Library of Congress)*

her friend "a mouth full of new years kisses."

In the autumn of 1840, Mary supplied Louis Agassiz with fossil fish that would help him and Egerton map the Rhaetic Bed, a layer of older rock under the lias. In this way, Mary helped geologists see which ocean creatures lived millions of years ago, long before the time of the dinosaurs.

Throughout her life, Mary Anning remained interested in all sorts of new ideas. In a notebook she kept during in the 1840s, she copied pages and pages of information about physics and astronomy. Reptiles that swam prehistoric seas fascinated Mary, but so did sunbeams, magnets, and distant planets.

After the death of her mother, Molly, in 1842, Mary lived alone for the first time in her life. Disappointed in love, she wrote or copied a poem called "To a Bride," which began: "The more divinely beautiful thou art / Lady! of Love's inconstancy beware."

She also jotted down a page of her thoughts for a poem she was composing:

> . . . and those who have loved the most
> too soon forget they loved at all.
> And such the change the heart displays
> So frail is early friendship's reign . . .

Mary copied poems in her notebook written by Henry Kirke White, Josiah Conder, and Lord Byron. On another page, she copied an essay about women's equal-ity—at a time when few people believed in the rights of women. Rejecting the way some people used the Bible to oppress women, the essay asked: "And what is a woman? Was she not made of the same flesh and blood as lordly Man? Yes, and was destined doubtless, to become his friend, his helpmate on his pilgrimage but surely not his slave, for is not reason hers? . . . Say then shall woman sink beneath the scorn of haughty man? No let her claim, the hand of fellowship . . ." In the 1840s, many people viewed religion as a set of rules they had to follow to avoid God's punishment. Mary approached faith differently. It helped her do uncom-mon work and triumph over difficulties.

Near the end of her notebook, when Mary was ill, she copied many prayers by Thomas Wilson, bishop of Sodor and Man. His "A Prayer under Lingering Illness" explored being "acquainted with grief," as Mary was herself, and longing "to see love as well as justice in all thy dealing . . . Make me so sensible of thy kindness and love that I may not only be contented but thankful under thy hand." She greeted the day with one of Wilson's morning prayers: "What shall I render unto the Lord for his mercies received unto me every morning? I will offer the sacrifice of thanksgiving and pay my vows unto the Most Highest. And may God accept my most hearty thanks for my preservation and refreshment, for all the blessings of the night past and of my life past."

Without her mother, Mary Anning tended her now-famous Fossil Depot by herself. In 1843, she offered two small ichthyosaurs to Adam Sedgewick at Cambridge. "Mr. Conybeare is quite in raptures with them," she told Sedgewick. One was "as perfect as if just taken from a dissecting room." The specimen was so well preserved that its intestines could be seen. She included what she called a "rough scratch," a drawing she made full of exquisite detail. Sedgewick bought both specimens.

Fossil shoppers kept coming from far and near. The following year an American geologist, Thomas B. Wilson, bought an ichthyosaur and a plesiosaur that he gave to the Academy of Natural Sciences of Philadelphia. In 1844, King Frederick Augustus of Saxony (now part of Germany), came to Lyme with his doctor, Carl Gustav Carus, who later helped develop the idea of species evolution. Carus recorded their visit to "a shop in which the most remarkable petrifactions and fossil remains—the head of an ichthyosaurus, beautiful ammonites, etc.—were exhibited in the window." The king paid Mary fifteen pounds for a six-foot-long skeleton of a baby ichthyosaur for his natural history collection in Dresden—and then asked for her autograph. After signing her name, she wrote, "I am well known throughout the whole of Europe."

That same year a terrible fire swept through Lyme. Forty buildings were destroyed, but once again Mary and her home were spared. The next month she wrote to her friend Dorothea Solly about the relationship between "Creatures of the former and Present World." Fifteen years before Darwin published his theory of evolution, Anning understood that species were not fixed and that ancient forms might be the direct ancestors of modern animals.

In her last years, Mary enjoyed scientific recognition, financial security, and a deepening religious faith. She could look back over her life with satisfaction. Though poor, she had formed friendships with wealthy collectors and geologists. Uneducated, she had aided the development of paleontology.

Mary still faced challenges in her last years, however. She learned by 1845 that she had breast cancer. Without a cure, doctors offered Mary laudanum (a mixture of opium and alcohol) to relieve the pain. The members of the Geological Society of London took up a collection for her, and the new Dorset County Museum elected her their first honorary member. Gestures such as these reassured her that she was not forgotten as she struggled with illness.

Mary remained alert to the end. Late in 1846, or early in 1847, she copied—or wrote herself—two long, humorous poems. "The Complaint of a Sunbeam against Dr. Faraday," described Michael Faraday's recent experiments in London with magnetism and light from the viewpoint of an abused sunbeam. The other was a limerick celebrating Roderick Murchison's geological accomplishments and the knighthood bestowed on him in 1846. The poem poked fun at Adam Sedgewick for clinging to old theories of the earth's formation, despite evidence Mary and others had found:

> Let Sedgewick say how things began
> Defend the old Creation plan
> And smash the new one,—if he can

Mary Anning died on March 9, 1847. She was forty-seven years old.

Even though she was never allowed to join the Geological Society of London, Henry de la Beche (who became its president) honored Mary with a speech which summed up her life and work, the only obituary ever given for someone who was not a member.

Fred Parry Hodges, the vicar of St. Michael's, along with members of the Geological Society, commissioned a portrait, painted by B.J.M. Donne, an eighteen-year-old Lyme Regis artist. They also installed a stained glass window at St. Michael's to memorialize Mary. Its six panels portray the "acts of mercy" Mary often performed, such as visiting the sick and feeding the hungry. It reads, "in commemoration of her usefulness in furthering the science of geology . . . her benevolence of heart, and integrity of life."

Appendix

The Fossils of Mary Anning

All the fossils Mary Anning sold to George Featherstonhaugh were lost when the Lyceum of Natural History in New York City burned to the ground. Several of Mary's finds, such as the "slender fin and slender jaw" ichthyosaur she discovered in 1824, were destroyed, along with most of the Bristol Institution, in World War II bombing raids, but other specimens have survived.

The skull of the first ichthyosaur Mary and Joseph discovered (*Temnodontosaurus platyodon*: "the cutting-tooth lizard with flat teeth") can be seen at the Museum of Natural History in London (earlier called the British Museum), along with the plesiosaur she found in 1823, her *Dimorphodon*, and part of the enormous ichthyosaur she discovered in 1832. William Gray's portrait of her, painted around 1842, welcomes visitors to the marine reptile gallery. An actress dressed as Mary Anning regularly performs there, telling visitors about her life.

The Lyme Regis Museum has an exhibit about Mary and presents lectures on her life each summer. Several of her other ichthyosaurs, a beautiful pentacrinite and the extremely rare head from a young Temnodontosaurus, can be seen at the Sedgewick Museum in Cambridge. One of the best stone lilies Mary ever found, a *Pentacrinus briareus*, and a particularly fine *Ichthyosaurus communis*, are at the Bath Royal Literary and Scientific Institution. The ammonite called Oxynoticeras that Mary found in 1838 is at the Somerset County Museum in Taunton. The National Museum of Natural History in Paris has a *Plesiosaurus dolichodeirus* Mary uncovered. One of her ichthyosaurs is at the Ulster Museum in Belfast, Northern Ireland, and the ichthyosaur and the plesiosaur Thomas Wilson bought are displayed at the Academy of Natural Sciences of Philadelphia, though it is not certain which specimens are hers.

As the British Museum of Natural History now says at the entrance to its Marine Reptile Gallery, "Some of the finest fossils in this gallery were found by Mary Anning . . . Fossil hunting became a life long passion and Mary Anning earned respect from collectors and scientists alike . . . and her remarkable fossils are still studied today."

Timeline

1799—Mary Anning is born on May 21.

1800—Mary survives lightning strike.

1807—Richard Anning is hurt in a fall.

1810—Richard dies.

1811—Joseph Anning finds ichthyosaur skull.

1812—Mary finds ichthyosaur skeleton.

1820—Colonel Birch has auction.

1821—De la Beche and Conybeare describe plesiosaur from Anning fragments.

1823—Mary finds complete plesiosaur.

1825—Mary meets the Murchisons.

1826—Mary buys her house.

1826—Featherstonhaugh buys fossils for New York Lyceum.

1828—Mary finds *Dimorphodon*.

1829—Mary finds *Squaloraja*.

1830—Mary finds a new plesiosaur.

1832—Mary & Joseph campaign for William Pinney.

1833—Mary narrowly escapes death twice.

1834—Mary works with Agassiz.

1835—Mary loses her life savings.

1839—Mary writes to the *Magazine of Natural History*.

1842—Molly Anning dies.

1844—King of Saxony visits Mary.

1845—Mary learns she has cancer.

1847—Mary Anning dies on March 9.

Sources

CHAPTER ONE: The Girl on the Cliff

p. 16, "a history and a mystery" *Chambers's Journal* 7 (1857): 314.

p. 17, "Martha Lock of Lyme Regis . . ." James Wheaton, *Theological Magazine and Review* 1 (1801): 35-37.

CHAPTER TWO: A New Vocation

p. 22, "All this land . . ." Jack Bowditch, quoted by John Fowles in his foreword to Elaine Frank, *The Undercliff: a Naturalist's Sketchbooks of the Devon to Dorset Coast* (Boston: Little, Brown, 1989), 9.

CHAPTER THREE: The Fish Lizard

p. 33, "I am going to sell . . ." Hugh S. Torrens, *Newsletter of the Geological Curators Group* 2 (1979): 409.

p. 35, "pick without hesitation . . ." Deborah Cadbury, *Terrible Lizard* (New York: Henry Holt, 2001), 101.

p. 35, "As I am a widow woman . . ." W.D. Ian Rolfe, Angela C. Milner, and F.G. Hay, *Special Papers in Palaeontology* 40 (1988): 148-149.

p. 36, "having been born a woman . . ." Harriet Blodgett, *Centuries of Female Days: Englishwomen's Private Diaries* (New Brunswick, NJ: Rutgers Univ. Press, 1988), 107.

p. 36, "low situation in life" Anna Maria Pinney, diary entry for January 23, 1832, at the Bristol University Library.

p. 36, "suddenly saw these hopes blasted" Ibid.

p. 37, "Sometimes, you know, after a storm . . ." Master
Hupjohn quoted in John Vaughan, *Monthly Packet* no. 4
vol. 6 (September 1893): 275-276.

p. 38, "the persevering industry . . ." George Cumberland,
Bristol Mirror 11(January 1823): 4.

CHAPTER FOUR: The Monster on the Beach

p. 39, "the magnificent specimen . . . " Hugh S. Torrens,
British Journal for the History of Science 28 (1995): 263.

p. 40, "like a snake pulled through a turtle" Samuel Wendell
Williston, *Water Reptiles of the Past and Present*
(Chicago: Univ. of Chicago Press, 1914), 77.

p. 40, "the most amazing creature . . ." Georges Cuvier,
Recherches sur les ossemens fossiles, vol. 5, part 2, (Paris:
Dufour & D'Ocagne, 1824), 476f.

p. 40, "a price no fossil . . ." Felix Farley, *Bristol Journal*
(March 6, 1824): 3.

p. 40, "the scientific public" William Daniel Conybeare, *Transactions of the Geological Society* 2nd series, vol. 1 (1824):
381, 389.

p. 41, "the well-known fossilist . . ." *New Monthly Magazine*
12 (February 1, 1824): 92-93.

p. 43, "What a leap we have here . . ." K.M. Lyell, *Life,
Letters, and Journals of Sir Charles Lyell* 1 (London: John
Murray, 1881), 151.

p. 42, "A link between ichthyosaurs and crocodiles," Henry
de la Beche & William Daniel Conybeare, *Transactions
of the Geological Society* 5 (1821): 559.

p. 44, "Mary Anning's knowledge . . ." W.D. Lang, *Dorset
Natural History and Archaeological Society Proceedings*
60 (1939): 153-154.

p. 45, "a magnificent specimen . . ." Lyell, *Life, Letters*, 153.

p. 45, "the famous fossilist . . ." Edwin Welch, *Devon and
Cornwall Notes and Queries* 32 (1973): 265-266.

CHAPTER FIVE: The Old Fossil Depot

p. 47, "A tempest teeming . . ." Sherborne & Yeovil, *Mercury*,
November 29, 1824.

p. 47, "it is quite a miracle . . ." letter from the Lyme Regis Museum's permanent exhibit on Mary Anning, May 1999.

p. 48, "to enjoy the sea air . . ." Archibald Geikie, *Life of Sir Roderick I. Murchison* 2 (London: John Murray, 1875), 334.

p. 48, "Our pocket money . . ." "S.E." (Helena Emma Waring), *Peeps into an Old Playground* (Lyme Regis: Dunster, 1895), 4.

p. 50, "very superb skeleton," Mary Anning to Gideon Mantell, November 24, 1825, quoted by Michael A. Taylor and Hugh S. Torrens, "Saleswoman to a New Science: Mary Anning and the Fossil Fish Squaloraja," *Dorset Natural History and Archaeological Society Proceedings* 108 (1986): 135-148.

p. 51, "a being of imagination . . ." Pinney, October 25, 1831.

p. 51, "as perfect as if . . ." David Price, *The Geological Curator* vol. 4, no. 6 (July 1986): 321-324.

p. 53, "very clever, funny Creature" E.B. Berkeley & D.S. Berkeley, *George William Featherstonhaugh, the First U.S. Government Geologist* (Tuscaloosa: Univ. of Alabama Press, 1988), 63, 66, 68.

p. 54, "I do enjoy an opposition among the bigwigs," Mary Anning to Charlotte Murchison, probably written at the end of 1828, quoted by Lang, *Dorset Proceedings* 76 (1956): 150.

p. 55, "sudden entombment" William Buckland, *Transactions of the Geological Society of London* 2nd series, vol. 3 (1835): 230.

CHAPTER SIX: The Flying Dragon and the Winged Fish
p. 56, "beating bits of green Sand . . ." letter from the Lyme Regis Museum.

p. 56, "an unrivaled specimen of Dapedium politum . . ." *Salisbury & Winchester Journal* 108 (December 1, 1828).

p. 56, "Of all the ancient beings . . ." Georges Cuvier quoted by William Buckland, *Geology and Mineralogy Considered with Reference to Natural Theology* (London: William Pickering, 1836), 273.

p. 57, "unknown species" William Buckland, *Transactions of the Geological Society of London*, 2nd Series, vol. 3 (1829): 217-218.

p. 58, "the lovers of science . . ." *Salisbury & Winchester* (March 2, 1829, and May 4, 1829).

p. 59, "industry and skill . . . " George Cumberland, *Quarterly Journal of Literature, Science, and the Arts* 27 (1829): 348.

p. 59, "I have never been out of the smoke of Lyme" Lang, *Dorset Proceedings* 66 (1945): 170.

p. 59, "with which I was much delighted" Anning's diary quoted by Lang, *Dorset Proceedings* 60 (1939): 160-161.

p. 59, "It is about a foot and a half . . ." *Salisbury & Winchester* (December 28, 1829).

p. 60, "It is quite unique . . ." Mary Anning to Adam Sedgwick, dated February 11, 1831, quoted by Michael A. Taylor & Hugh S. Torrens, *Dorset Natural History and Archaeological Society Proceedings* 108 (1986): 136, 142.

p. 60, "It is to her almost exclusively . . ." Edward Pidgeon, *The Fossil Remains of the Animal Kingdom* (London: Whittaker, 1830), 377.

p. 63, "last week I discovered . . ." Lang, *Dorset Proceedings*, 60 (1939): 155-156.

p. 63, "When we see . . ." Buckland, *Geology and Mineralogy*, 201-202.

p. 64, "Approach, approach, ingeneous youth . . ." Nicolaas A. Rupke, *The Great Chain of History: William Buckland and the English School of Geology 1814-1849* (Oxford: Clarendon Press, 1983), 142.

p. 64, "noticed by all the cleverest men . . ." Pinney, October 25, 1831.

p. 64, "The tide warns me . . ." Taylor and Torrens, *Dorset Proceedings,* 136.

p. 64, "the best she had ever seen . . ." William Buckland to H.T. de la Beche, May 1, 1831, at the National Museum of Wales, NHMS XXXVII/17f.

p. 65, "those men of learning . . ." Pinney, October 25, 1831.

p. 65, "I beg your pardon for distrusting . . ."*All the Year Round* 18 (1865): 62.

CHAPTER SEVEN: The Lioness of Lyme Regis

p. 66, "Before I knew what to do . . ." Pinney, October 25, 1831.

p. 67, "I beg ten thousand pardons . . ." Taylor and Torrens, *Dorset Proceedings,* 135f.

p. 68, "had no sweethearts . . ." Pinney, January 23, 1832.

p. 68, "this extraordinary character . . ." Pinney, October 25, 1831.

p. 68, "I can enjoy everything but gloomy silence . . ." Anning's diary, quoted in the Woodword Collection of the Blacker-Wood Library at McGill University in Montreal.

p. 68, "the wild romance . . . " Pinney, January 23, 1832.

p. 68, "She has more than the power . . ." Pinney, October 25, 1831.

p. 69, "She is very kind . . ." Pinney, January 23, 1832.

p. 70, "It is certainly uncommon . . ." Maria Hack, *Geological Sketches and Glimpses of Ancient Earth,* (London: Harvey & Darton, 1832), 302.

p. 71, "We sallied out . . ." Gideon Mantell, *The Journal of Gideon Mantell, Surgeon and Geologist,* edited by E. Cecil Curwen (London: Oxford University Press, 1940), 108.

p. 72, "the best yet discovered" Price, *Geological Curator,* 320-321.

CHAPTER EIGHT: Praise

p. 75, "like a sword . . ." Pinney, August 21, 1832.

p. 76, "the death of my old faithful dog . . ." Lang, *Dorset Proceedings* 66 (1945): 169-173 and 71 (1950): 184-188.

p. 76, "Yesterday she had one of her miracle escapes . . ." letter dated December 9, 1833, in the Oxford Univ. Museum of Natural History, Buckland letters, box 2, 7.

p. 76, "The Word of God . . ." Pinney, December 8, 1833.

p. 77, "Associating with . . ." Pinney, October 25, 1831.

p. 78, "the plesiosaurs and pterodactyls . . ." William Whewell, *British Critic* 4th series, vol. 9, (1831): 194.

p. 79, "among geologists . . ." John Murray, *The Truth of Revelation* (London: Longman, Murray, 1831), 89.

p. 79, "The plesiosaurus somewhat resembles . . ." Ibid., 99.

p. 79, "the truth of revelation" Ibid., 169.

p. 82, "already well-known in every part of the world . . ." George Roberts, *The History and Antiquities of the Borough of Lyme Regis and Charmouth* (London: Samuel Bagster, 1834), 128.

p. 82, "genius for discovering . . ." Ibid., 284.

p. 83, "This lady, devoting herself to Science . . ." Thomas Hawkins, *Memoirs of Ichthyosauri and Plesiosauri* (London: Relfe & Fletcher, 1834), 9, 26.

p. 83, "the smallest I have seen . . ." Price, *Geological Curator*, 321.

p. 83, "the tide would not allow . . ." Ibid.

p. 84, "zeal and intelligence" Ibid.

p. 84, "Miss Mary Anning, to whom the scientific world . . ." Buckland, *Geology and Mineralogy* 1, 198, 304, 437-438.

p. 85, "No one ought to go near . . ." Samuel Goodrich, *Peter Paley's Wonder's of Earth, Sea and Sky* (New York: Colman, 1837), 7-17.

p. 85, "We had the pleasure of making . . ." M. Aurousseau, *The Letters of F.W. Ludwig Leichhardt* 1 (London: Cambridge Univ. Press, 1968), 54, 62-63.

CHAPTER NINE: Through the Storm

p. 86, "Though keenest winds were whistling round . . ." John Kenyon, *Poems, for the most part occasional* (London: Edward Moxon, 1838), 109-111.

p. 87, "the most perfect jaw . . ." Edward Charlesworth, *Magazine of Natural History*, New Series, vol. 3 (April 1839): 243

p. 87, "were the teeth by which the fish . . ." Anning's letter dated April 7, 1839, *Magazine of Natural History* (December 1839): 605.

p. 87, "as I am illiterate . . ." Mary Anning to Edward Charlesworth, July 12, 1839, in the Blacker Wood Library of Zoology and Ornithology at McGill Univ.

p. 88, "although I am almost . . ." Mary Anning to Miss M. Lister, January 13, 1840, quoted by Lang, *Dorset Proceedings* 71 (1950): 184-185.

p. 90, "The more divinely . . ." Anning's "Fourth Notebook," manuscript XXXVII/2 from the Lang papers at the Dorset County Museum.

p. 91, "And what is a woman . . ." Anning's diary from the Blacker Wood Library.

p. 92, "Mr. Conybeare is quite . . ." Price, *Geological Curator*, 323.

p. 92, "I am well known . . ." C.G. Carus, *The King of Saxony's Journey Through England and Scotland in the Year 1844* (London: Chapman & Hall, 1846), 197.

Glossary

ammonite: an extinct relative of the modern chambered nautilus, with a pointed head, two large eyes, and muscular tentacles around its mouth. Most grew shells coiled into a tight, flat spiral, but a few shells were shaped like spears or corkscrews.

belemnite: a prehistoric relative of squid and cuttlefish, as large as five or six feet long, with hooks on their tentacles rather than suction cups.

coprolite: fossilized feces sometimes found inside a fossilized animal's intestines or expelled in death. These "dung-stones" reveal the diet and the shape of the stomachs and intestines of ancient creatures.

crinoid: a flower-shaped animal, still surviving today, related to starfish, sea urchins, and sand dollars, which uses its feathery "arms" to strain tiny food particles from sea water. Most "sea lilies" have a five-sided stem that anchors them to rocks or floating pieces of wood as they grow. They break loose from these moorings as adults, swimming freely and trailing the long stems behind them.

***Dapedius*:** a swift, beautiful fish about a foot long. Its short mouth, peg-like teeth, and strong jaws could break the hard shell of a belemnite.

genus: a classification of living creatures more broad than species. The species *Homo neanderthalensis* (cave-dwelling Neanderthals) and our species, *Homo sapiens* (modern humans), are in the same genus, *Homo,* or human.

Hybodus: the earliest known shark. A highly successful predator, it was covered by plates, had a strong skull, and spines in front of each fin on its back (the ichthyoduralites or "fish spears" Mary discovered). Males had horn-like spines sticking out of their heads.

ichthyosaur: an ancient swimming reptile resembling a dolphin but not closely related to it or any other reptiles. Unlike modern reptiles, ichthyosaurs gave birth to live babies. Ichthyosaurs were the biggest, most widespread predators of their time. They are the only prehistoric reptiles for which the skin color is known: its back was tortoise-shell brown.

ophiuroids: these "brittle stars" or "basket stars" are rarely preserved as fossils (because they usually fall apart after they die) but still live in every ocean. These creatures whip their five long, slender, flexible arms to move through water, giving them the common name, "snake stars."

pterosaur: the first creature with backbones to fly, these flying reptiles dominated the skies for 135 million years and evolved into the largest flying creatures the world has ever seen. Many scientists believe they were warm-blooded and covered with fur.

plesiosaur: these swimming reptiles flourished in the seas alongside ichthyosaurs. Powerful predators, they may have moved like seals and sea turtles, "flying" underwater.

species: creatures that share many characteristics and can produce fertile offspring.

Squaloraja: a primitive, extinct chimaera fish, related to sharks and rays. Its diet consisted of mostly shellfish.

Selected Bibliography

Alic, Margaret. *Hypatia's Heritage: A History of Women in Science from Antiquity through the Nineteenth Century*, Boston: Beacon Press, 1986.

Cadbury, Deborah. *Terrible Lizard: The First Dinosaur Hunters and the Birth of a New Science*, New York: Henry Holt, 2001.

Fowles, John. *The French Lieutenant's Woman*, Boston: Little, Brown, 1969.

————.*A Short History of Lyme Regis*, Boston: Little, Brown, 1982.

Goodhue, Thomas W. "The Faith of a Fossilist: Mary Anning," *Anglican and Episcopal History* 70 (March 2001): 80-100.

Gould, Stephen Jay, and Rosamund Wolf Purcell. *Finders, Keepers: Eight Collectors*, New York: Norton, 1992.

McGowan, Chris. *The Dragon Seekers: How an Extraordinary Circle of Fossilists Discovered the Dinosaurs and Paved the Way for Darwin*, Cambridge, MA: Perseus, 2001.

McMurtry, Jo. *Victorian Life and Victorian Fiction*, North Haven, CT: Shoe String Press, 1996.

Ogilvie, Marilyn Bailey. *Women in Science: Antiquity through the Nineteenth Century*, Cambridge, MA: MIT Press, 1986.

Pool, Daniel. *What Jane Austen Ate and Charles Dickens Knew: From Fox Hunting to Whist—the Facts of Daily Life in Nineteenth Century England*, New York: Simon & Schuster, 1993.

Torrens, Hugh S. "Mary Anning (1799-1847) of Lyme; `the greatest fossilist the world ever knew,' " *British Journal for the History of Science* 28 (1995): 257-284.

Websites

Lyme Regis Museum: Mary Anning and the Birth of Geology
http://www.lymeregismuseum.co.uk/fossils.htm

Southhampton University: Geology of the Dorset Coast
http://www.soton.ac.uk/~imw/lyme.htm

University of Califonia at Berkeley: "The greatest fossilist the world ever knew"
http://www.ucmp.berkeley.edu/history/anning.html

University of California at Berkeley: Links to dinosaur pages on the internet
http://www.ucmp.berkeley.edu/diapsids/dinolinks.html

Zoom dinosaurs: A comprehensive guide to dinosaurs for students of all ages
http://www.enchantedlearning.com/subjects/dinosaurs/

Index